'Maths is just another "language" – and Eddie Woo makes it so easy and fun.'

Dr Karl Kruszelnicki, science commentator, author and Julius Sumner Miller Fellow in the Science Foundation for Physics

'For a mathematician, Eddie Woo is one helluva storyteller. An excellent and important read from beginning to end.'

Maxine McKew, Honorary Enterprise Professor, University of Melbourne

'Enthusiastic, energetic Eddie Woo explores mathematics in ways that reveal how human and beautiful it is.'

Nalini Joshi, mathematician, University of Sydney

'You probably know acclaimed math teacher Eddie Woo through watching his excellent videos on his Wootube channel. Well, now there is a book, and it's a winner. A compendium of short essays where Mr Woo shows how mathematics lies just beneath the surface in practically every aspect of our lives. What makes it sing is that his engaging personality shines through on every page, just as much as it does on video when he is in front of a class.'

Keith Devlin, Stanford University mathematician and author of many popular mathematics books

'Eddie Woo is an inspirational maths teacher. Why? Because he can also communicate, connect and write.'

Jane Caro, social commentator and bestselling author of *Plain-speaking Jane*

'Learning mathematics is like climbing the stairs of a skyscraper. It's difficult and can seem utterly pointless. Some educators scream at us from a 10th-storey window as we look up at them in confusion. Eddie greets you at the foyer and is there beside you while you take each and every step. And once at the observation deck, he's admiring the beautiful vista with you. Eddie is more than just the maths teacher we all wanted. Eddie is the maths teacher we all need.'

Simon Pampena, Australian Numeracy Ambassador, Numberphile Star

'Eddie Woo's gift is in using stories to help us see the way maths breathes life, colour, shape, and rhythm into the world around us. He's transformed the lives of countless students in his classroom and on Wootube. Now he's here to change how you see numbers too – whether you think you have a mathematical mind or not!'

Natasha Mitchell, science journalist and radio host

'A sweeping tour de force of how to engage people with mathematics.'

Matt Parker, comedian, author and maths communicator

WOO'S
WONDERFUL
WORLD
OF MATHS

First published 2018 in Macmillan by Pan Macmillan Australia Pty Ltd
1 Market Street, Sydney, New South Wales, Australia, 2000

Reprinted 2018 (six times), 2019 (twice), 2020

Cataloguing-in-Publication entry is available
from the National Library of Australia
http://catalogue.nla.gov.au

Cover and text design by Alissa Dinallo
Illustrations by Hannah Schubert and Alissa Dinallo
Author photographs by Alissa Dinallo
Typeset in Electra LH by Alissa Dinallo

Internal images courtesy Shutterstock; photograph on page 39 courtesy Michael Taylor, Landsat Project Science Office, NASA Earth Observatory; image on page 143 courtesy iStock; graph on page 338 courtesy of Proceedings of the Royal Society of London. Series B, Biological sciences; graph on page 345 courtesy Shane Oliver, AMP Capital.

Printed by IVE

WOO'S WONDERFUL WORLD OF MATHS

Eddie Woo

Pan Macmillan Australia

Dedicated to the Author of Life

'Mathematics is the language with
which God has written the universe'
— Galileo Galilei

About

EDDIE WOO

Eddie Woo is the head mathematics teacher at Cherrybrook Technology High School, Sydney. He has been teaching mathematics for more than ten years.

In 2012, Eddie started recording his lessons and uploading them to YouTube – creating 'Wootube'. Since then, he has amassed a following of more than 300,000 subscribers and his videos have been viewed more than 18 million times.

In 2018, Eddie was named Australia's Local Hero of the Year and shortlisted as one of the top ten teachers in the world.

CONTENTS CONTENTS CONTENTS

CONTENTS CONTENTS CONTENT

PROLOGUE

When I was a school student, I found little joy in mathematics. I understood parts of it, but always found it discouraging: doing maths felt like trying to memorise an arbitrary set of rules in a game I didn't understand and didn't even really have any interest in winning. Despite being able to wrap my head around quite a few of the ideas and theorems, I seldom experienced success because I kept making what my teachers always called 'silly mistakes' – careless errors and flaws in the accuracy of my calculations that prevented me from getting the right answer.

As a teenager, that seemed to be what maths was all about to me: learning ways to take a problem and find some elusive number trapped within it, the 'solution'. Since I never found it very easy to do that, I tolerated mathematics but never enjoyed it or felt like I was any good at it. Instead, I focused on subjects that came to me much more naturally: English, history and drama. But all that began to change the year I turned nineteen.

I sincerely hope that many of you opening this book share a story like mine. Mathematics was never your thing. The reason I hope this is because if you're holding this book and are about to dive into its pages, then like me at age 19, your story isn't over yet. Because you see, when I was 19 years old I began training to become a mathematics teacher. That may sound a little surprising to you given the way I described myself earlier – and I promise I'll explain how I ended up in this unusual position! But what matters for now is this: as I started out on the road to becoming a high school teacher, I learned a secret. Actually, I learned hundreds of them – because I began to discover that mathematics was something very different from what I had thought it was. I started to uncover what Polish mathematician Stefan Banach was talking about when he said that 'mathematics is the most beautiful and most powerful creation of the human spirit'.

That's what this book is about. I want to take you with me on the journey I went on to understand that

maths

is all

around

us.

Maths enables us to see and touch the invisible realities that make our universe what it is, and maths can help us more deeply appreciate all the things we love in this world. Those are some pretty big goals – so we best get to it!

Happy reading.

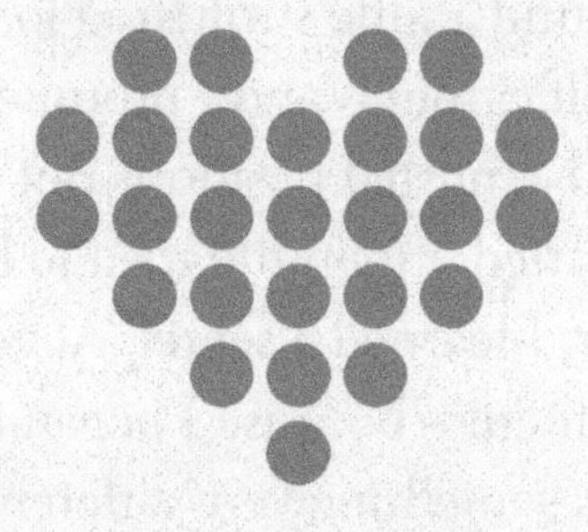

Eddie

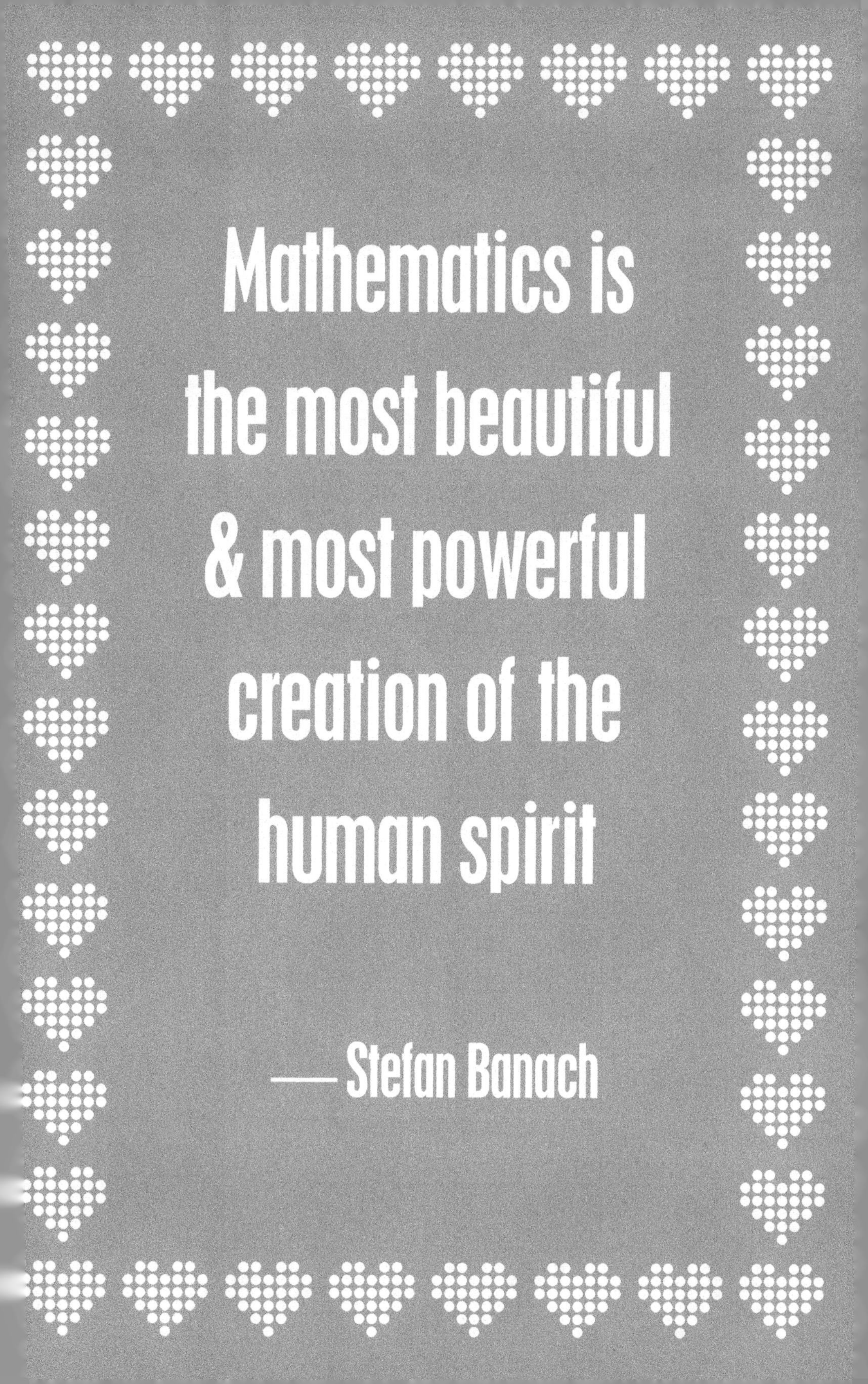
Mathematics is
the most beautiful
& most powerful
creation of the
human spirit
— Stefan Banach

CHAPTER 01

BORN MATHEMATICIANS

ARE HUMANS BORN MATHEMATICIANS?

This question was posed to me once during a radio interview. It came in the context of the assertion that humans are born scientists. You don't need to teach a child to experiment with their surroundings, observe the effects and then repeat the process until they can confirm or deny a hypothesis. This behaviour is instinctive and requires no formal training. In this way, even if they can't articulate it, children are thinking and working scientifically from the moment they first open their eyes and start to investigate the world around them.

So, are people born mathematicians? Do children think and work mathematically on their own or is this a learned behaviour?

One of the reasons this question is on my mind is because it links closely to an idea that many people hold, which is that some people are born with mathematical ability while others are not. It usually comes in the form of the personal admission,

'I'm not a maths person.'

It's fairly common for people to think of mathematics as a special talent that only some people possess. If you aren't born with it, you can never really get it. Many people say this of themselves (and teach it to their children!) – but does it have any basis in reality?

To settle on an answer to this, we really need to work out first what we mean by a mathematician. This turns out to be harder to define than you might initially think.
A biologist is someone who studies living things. A physicist studies moving things. A chemist studies substances. An astronomer studies stars and planets. A geologist studies rocks. These are all very well-defined fields with nice, neat boundaries. But what about a mathematician? What do they study? A knee-jerk response might be to say that a mathematician studies numbers, but there are entire fields of mathematics that can be explored quite deeply without discussing numbers (such as geometry or topology). So what is it that all mathematicians have in common?

The answer most people will agree on is that all mathematicians study patterns. A pair of odd numbers always add to an even number. The exterior angles of any polygon, no matter how big or small or irregular, always add up to a full revolution. The rows of Pascal's Triangle always add up to a power of 2.

What do all mathematicians have in common?

A
THEY
STUDY
PATTERNS

The path of an object under the force of gravity always traces out a curved shape called a conic section (either a circle, ellipse, parabola or hyperbola). The florets of a flower always spiral outwards according to a very specific (and ingenious) geometric pattern.

This is why it's impossible to put a fence around what mathematicians are interested in: they are interested in any kind of pattern, and patterns exist everywhere.

We live in a patterned universe, a cosmos.

That's what cosmos means (orderly and patterned) – as opposed to chaos (disorderly and lacking sensible patterns).

Now we can actually define the question we started with. When we ask, 'Are humans born mathematicians?', what we are really asking is, 'Are humans born to seek out and try to understand the patterns around us?'

Stating the question in this way makes things clear. The answer is emphatically yes. The human brain is nothing if not a pattern-recognising machine, built from the ground up to perceive patterns in our surroundings. You can describe virtually every function of the brain in terms of its relation to patterns. What is smell? It's our recognition of specific olfactory patterns and associating some of them with good (sweet) and some of them with bad (bitter). What is memory? It's the connection of patterns with specific meanings, like the facial and vocal cues of people we meet whom we can therefore later recognise.

Much of what we would describe as understanding or skill is the ability to recognise patterns more effectively than others. An experienced doctor can recognise a condition through a particular pattern of symptoms. An expert taxi driver knows the most efficient pattern of roads and turns to take to get to their destination given their current location and traffic conditions. And as we gain practice at performing certain patterns over time, they become a part of our character and personality – we call them habits.

It isn't just seeing patterns that we humans are so good at. We love making our own patterns, and the people who do this well have a special name – we call them artists. Musicians, sculptors, painters, cinematographers – all of them are creators of patterns, and hence they are also mathematicians in their own way. I once heard music described as 'the joy that people feel when they are counting but they don't know it'. Due to Islam's aversion towards representations of humans and animals, Islamic design primarily consists of intricate arrangements of tiles that are literally geometric patterns.

Humans are so accustomed to looking for patterns that we even see patterns where they don't exist. The gambler's fallacy and part of the placebo effect are perfect examples of our unstoppable desire to link cause and effect in our daily experience, even when careful logic actually suggests otherwise.

So, yes – I think humans are born mathematicians.

We aren't necessarily born as good ones! But that's why I love being a mathematics teacher, and it's what drives me to help people grasp this subject. When we grow as mathematicians, we become better at pursuing the deeply human drive to understand the beauty and logic behind the patterns that animate the universe.

CHAPTER 02

THE HEAVENLY CIRCLE

'Daddy, look out the window!'

I'm trying to concentrate on driving on this wet afternoon, and I'm squinting hard even through my sunglasses because the sun is so low on the horizon and the road is glistening with moisture. The school pick-up is a stressful operation at the best of times, but my daughter's voice catches my attention from the back seat and I lift my gaze to look at her in the rear-view mirror. Her elbow is on the door's armrest and she's supporting her chin with her hand as she stares through the rain-speckled window. I can see by the look in her eyes that something has transfixed her. So I turn and look, and there it is: the brightest rainbow I've seen in years. I look at it for longer than I should, given the fact that I'm still moving slowly through traffic, but like my daughter I find it hard not to stare. The iridescent green, the glowing red, the unearthly indigo . . . Though I've seen rainbows like this hundreds of times before, there's something about today's display that makes it particularly arresting.

'Why's it round, Daddy?'

'Mmmm?' I reply, in that way parents do when they are just too distracted to give a decent response. My eyes

lock back onto the road ahead of me and the traffic that's now stopped around us. My brain finally catches up, but I reflexively repeat the question to stall for time anyway. 'Round?' She's still looking out the window, but I can see her nod in my peripheral vision. 'Yeah, why's it round?'

There are so many things I love about my children. One of my favourite qualities is their perpetual wonder. Due to their age, or rather lack thereof, they have eyes to see things in the world – genuinely beautiful and amazing things – that I have become bored with, that my brain has trained itself to ignore.

Case in point: why is a rainbow a bow? What gives it its elegantly round shape?

It turns out that the elegant roundness of a rainbow comes from a surprising source: the elegant roundness of each and every raindrop that makes up the rainbow.

I say surprising because most people, ironically, do not picture raindrops as round. On the contrary, a cursory search on the internet for 'raindrop' will turn up millions of images that are distinctly pointy at the top. However, a search for 'raindrop photos' reveals a more realistic picture: while sometimes slightly stretched or squashed, raindrops are nonetheless much closer to spherical in shape than these caricatures would have us believe.

But we're getting ahead of ourselves. Let's rewind a little to think about what's happening when we see a rainbow in the sky. As we know from experience, rainbows don't always form after rain; the sun needs to be shining brightly enough soon after the rain ends for a bright rainbow to appear, which is why they are often seen around sun showers. If your entire sky is blanketed in thick clouds, then you're out of luck. Rain is necessary, but not sufficient – you really need the light.

Fans of Pink Floyd and Isaac Newton alike will know that light does a curious thing when it passes through something like a prism. Due to a phenomenon we call **refraction,** white light – like the kind emitted by our sun – splits up into its components, which we call a **rainbow.**

Raindrops behave a bit like the piece of the glass on the album cover, refracting sunlight and dispersing it into a spectrum. But if that was all that was going on, we'd expect

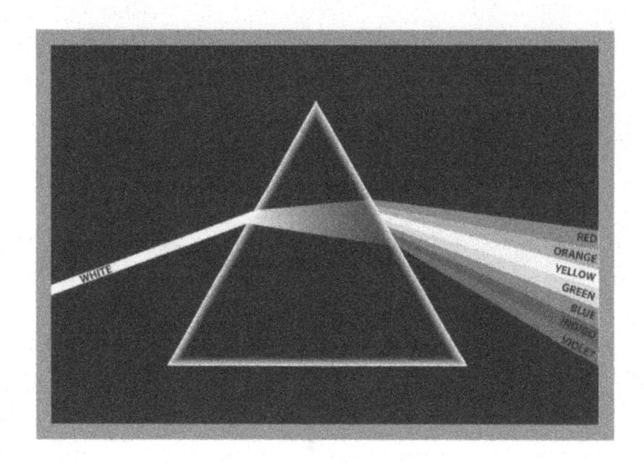

there to be rainbow light everywhere after a shower ends. Why is the rainbow bound up in a tight band that seems to belong in a perfect circle? And besides that, why does the rainbow curve away from the sun rather than around it?

That's because the geometry of a circle makes the rays of light coming from the sun behave in a very predictable – and literally dazzling – way when they interact with a sphere of water. They not only disperse into their component colours, but they also reflect inside the raindrop in such a way that they all shine back perfectly in one particular direction, revealing the whole spectrum of colours as they do so.

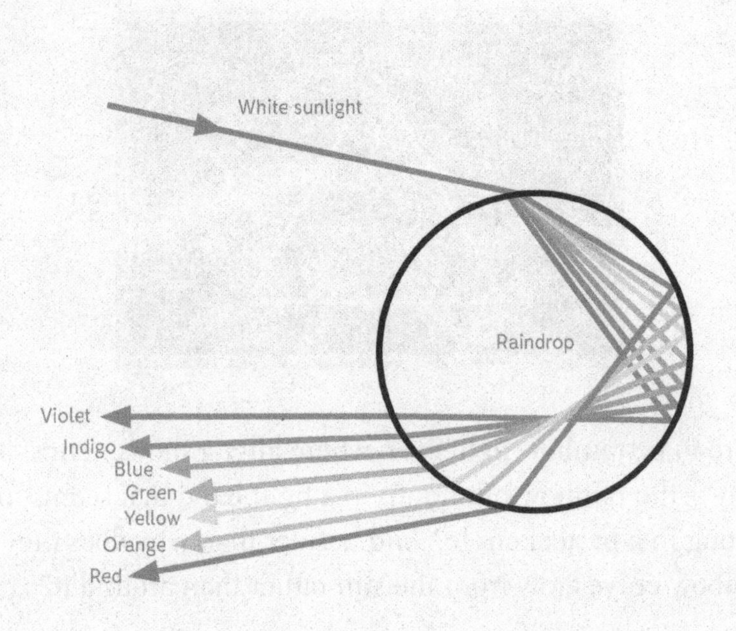

There are millions of other raindrops that also reflect light towards the precise position where you are standing – and every single one of those droplets lies on the surface of a gigantic cone with your eye at the tip. When you look at a cone from the perspective of the tip, though, you don't see the full cone – you just see the cone's cross-section, which is a circle. You might then ask why we only see semi-circular rainbows; the reason is that the horizon tends to obscure the bottom half of the circle. The full version can indeed be seen when viewed from the air, such as out the window of a plane if your timing is lucky enough!

For me, this is what mathematics is. The world around us is filled with patterns, structures, shapes and relationships that beg to be not just marvelled at but understood.

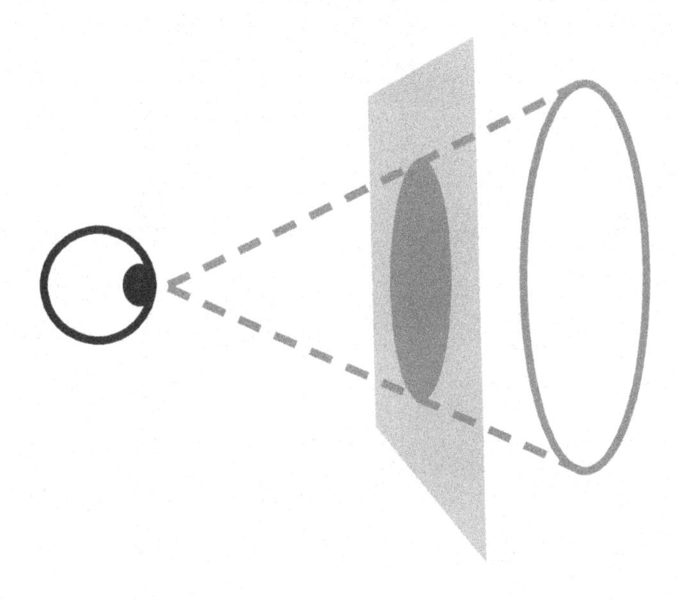

Humanity developed mathematics as a language we can use to interpret the world, but realities such as the rainbow have taught us that mathematics is no mere invention. It has always been woven into the fabric of the universe all around us, if we're willing to open our eyes and see it.

I don't remember how I answered my daughter that afternoon while we sat in traffic and admired the sky. But now I can tell her – and you – that rainbows are round because a choir of raindrops has conspired together to put on a lightshow so breathtaking and ethereal that if it didn't just appear in the sky above our heads, we probably wouldn't believe it.

CHAPTER 03

MUSIC TO MY EARS

The acoustic guitar sitting beside my desk is a marvellous piece of design. Every time I strum the strings, it's making the best-sounding mathematics you can imagine.

Humans have been making music ever since . . . well, ever since there have been humans. But it was Pythagoras – yes, that Pythagoras, the one best known for tormenting children around the world with right-angled triangles – who was said to have discovered and articulated the mathematics that gives us the musical notes we know and love.

As the story goes, Pythagoras was walking along when he passed a blacksmith. Inside the smithy, a pair of workers were hammering away on their anvils – and each anvil, being a different size, made a different sound when it was hit. It was at that moment that he realised there must be a mathematical relationship between the object's size and the sound that it made. Experimenting on some metal bars that the blacksmith had left near the street – because when you're a free-thinking ancient Greek philosopher, no one minds if you hit their belongings for no apparent reason – he noticed that it sounded especially good if you struck a bar together with one that was exactly half its length. To understand why, we need to know a little bit about how sound works.

Whenever air vibrates, our ears perceive it as sound. Anything that moves the air will make sound – your footsteps on the ground as you walk, the hundreds of gears and pistons in your car's engine when you drive, or the wind howling outside during a storm. We can represent those sounds visually using a graph that shows how much the air is vibrating in a particular spot as time passes. Here's what the graphs of footsteps, car engines and windy weather look like:

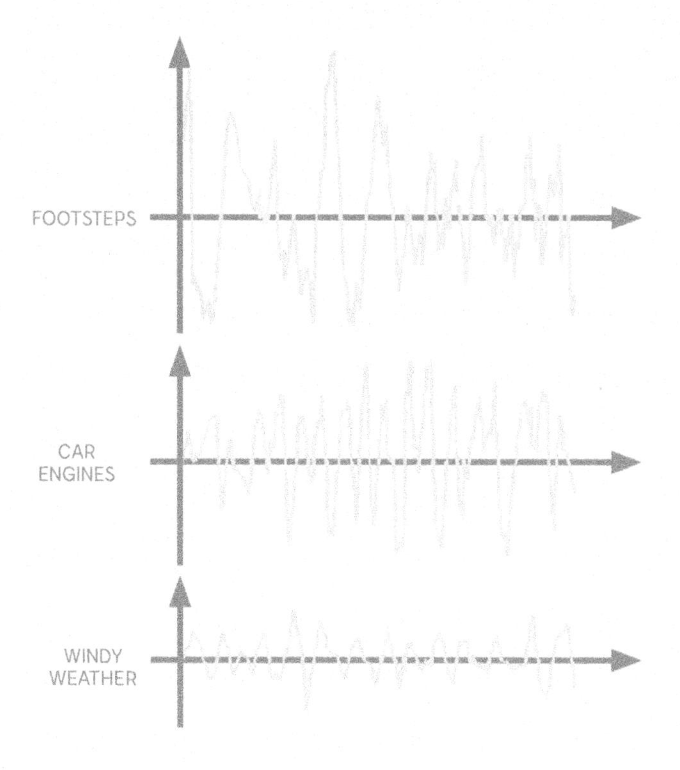

While these graphs don't seem to look very much like each other, they all share something in common. None of these graphs represents a musical sound, so their graphs wave up and down with what appears to be chaotic unpredictability. Compare them to the following graphs, which each represent a musical note:

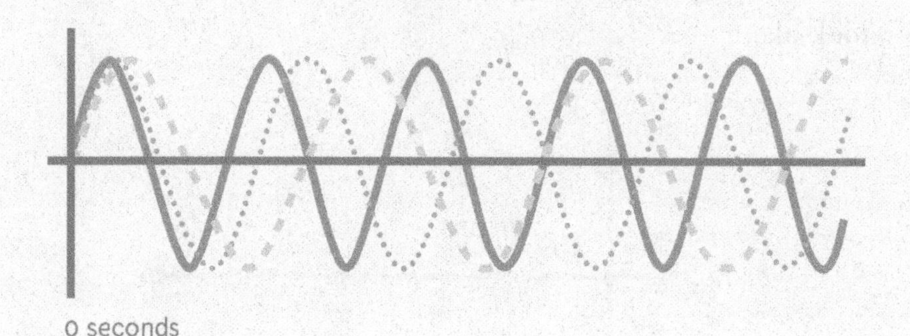

KEY

A

F#

D

The difference is immediately striking. Mathematicians call these kinds of graphs 'periodic', since they repeat over and over at consistent intervals of time ('periods'). The actual shape you're looking at gets the fancy title of 'sinusoidal wave', because 'sinus' comes from the Latin term for 'curve' (your sinuses are a series of 'curved' recesses). Mathematicians, famously lazy and always looking for ways to abbreviate things, call it 'sine' for short. A sine graph that waves up and down more quickly has a high frequency, which we hear as a higher pitch, while one that waves slowly has a low frequency and a low pitch.

Musical instruments can make these very simple sound graphs because they are actually very simple objects. For instance, the instrument I am most familiar with – the acoustic guitar – is basically just a set of strings that move up and down to vibrate the air around them. The hollow body of the guitar just provides a space where the vibrations can echo and amplify, producing a louder sound. But the actual essence of the guitar is just a string.

When you pluck a guitar string – or any string that is held tight enough, for that matter – it oscillates up and down, vibrating the air around it and creating a lovely note. But – and this is where we start to understand what Pythagoras discovered – the beauty of a musical instrument is that you can make it play different notes. On a guitar, the way you do that is by holding down a string on one of the vertical bars positioned on the guitar's neck, which are called frets.

Pressing down on a fret essentially makes the string act as though it were a shorter version of itself – the part that you pluck has a smaller length than if you had left the string untouched. And a shorter string can vibrate up and down faster, while a longer string has to vibrate up and down more slowly. An easier way of thinking about this is to imagine some children playing with a really long skipping rope, the kind that allows you to fit four or five people inside all at the same time. That rope will go more slowly than one which is only designed for a single person, and guitar strings are just the same.

This isn't the only thing you can do to control the speed of a string's movement; thicker, heavier strings also move more slowly and so they produce notes with lower pitch. That's one of the differences between the six strings you find on a guitar – the strings get progressively thicker to enable them to make deeper and deeper sounds. (This is also why, if you walk into a music store and compare a regular guitar to its bass counterpart, you'll find the bass has much thicker strings.)

Let's come back to Pythagoras. Suppose this was the set of metal bars he stumbled across, along with their lengths.

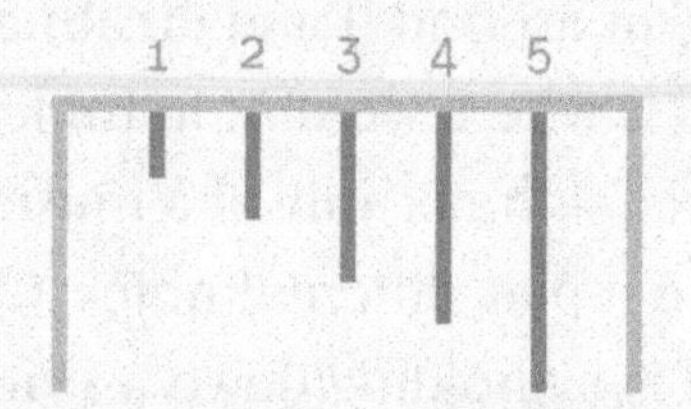

Bar 2, being exactly half the length of Bar 4, vibrates faster – twice as fast, in fact. So when you compare their sound graphs, they look like this:

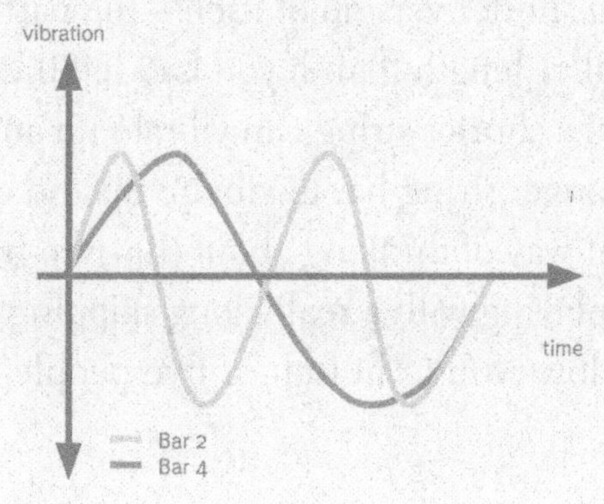

You can see that over and over again, the sounds start vibrating and stop vibrating together. And this is what we hear and feel as harmony: these notes sound great together. In particular, these two notes are what musicians call an octave.

Music is the art of taking these chords and combining them in ways that form an emotional journey. For instance, the consummate composer Beethoven is known for his astonishing skill in arranging sounds of consonance – like the harmony shown opposite – together with sounds of dissonance, which cause the listener to yearn for a resolution. Dissonant chords look very different when you consider their sound graphs:

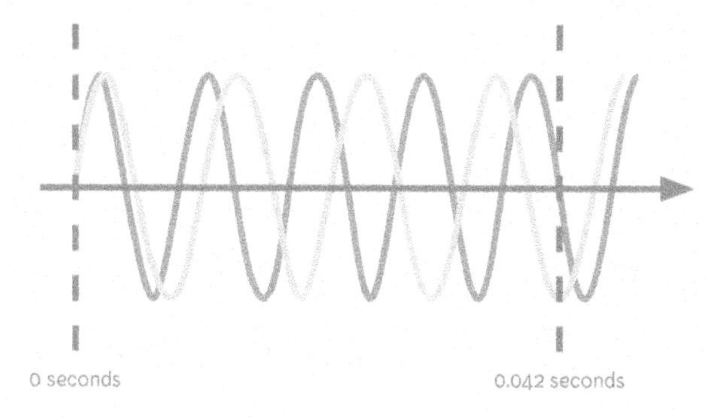

You can see here that the musical notes in dissonant chords get very close, but never seem to start or stop together at neat intervals. The human ear bristles at such sounds, which is why our musical yearning is actually the unconscious desire for mathematical harmony.

CHAPTER 04

LIGHTNING THROUGH YOUR VEINS

Poetry is the art of calling the same thing by different names. Mathematics is the art of calling different things by the same name.

— Henri Poincaré

We live in a universe of themes: concepts and principles, like love and magnetism, exist everywhere and sometimes turn up where you least expect them. And we are creatures who are built to seek connection: we love to join the dots, to understand how things are interrelated and appreciate the links between seemingly disparate ideas. In a time when it is all too easy to focus on differences, it is reassuring to know that so many things have a deep and profound unity lying just beneath the surface.

This is one of the reasons why I think that mathematics is beautiful. As Bertrand Russell put it, it's a beauty that is 'cold and austere'; that is to say, among other things, it can be something of an acquired taste. However, if we're willing to invest some effort in wrapping our heads around it, we are rewarded with a richer and clearer view of the wonderful world we live in. In particular, mathematics has this unique power to help us see how things that may appear very different are actually intimately connected. Often, things that seem to inhabit completely separate worlds are shaped and animated by the same principles. One of my favourite examples of this is blood vessels and lightning bolts.

On the face of it, no two things could be more different from one another. One is living; the other is inanimate (though it certainly has the power to end life!). One is as human as you are; the other is more akin to an act of the gods. One can be so fine that it is invisible to the human eye; the other can dwarf the largest skyscrapers. One is made of squishy flesh and pulsating liquids; the other is incorporeal energy in arguably one of its purest forms. Yet the two share an unmistakable bond, due to one of their most obvious features: their shape.

Admittedly, most of us don't spend a great deal of time looking at either of them. Lightning obviously doesn't like to stick around for very long, and there are often things obstructing your view so you can't appreciate it in its full glory. On the other hand, your body is literally filled with blood vessels, but most of them are buried deep in your muscles and hidden from view. So let me help you out by filling in some visual gaps for you:

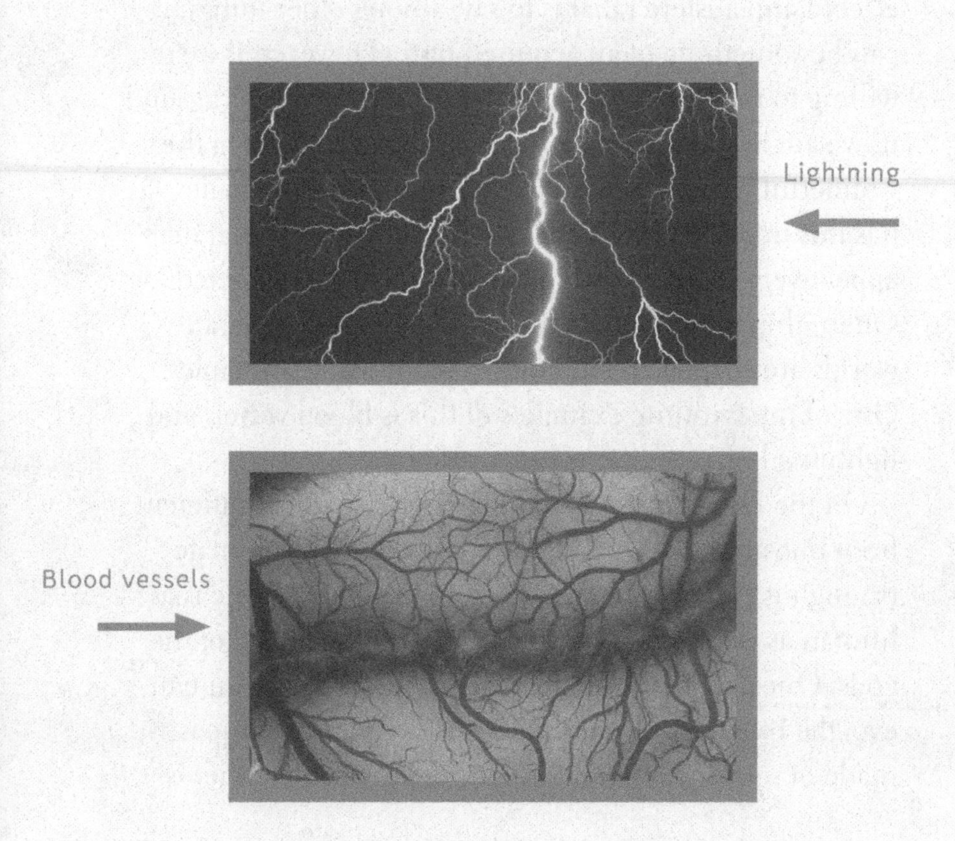

You might say that it's – pardon the pun – quite striking that such different objects should be so alike. The burning question is why? Is there a reason why such different objects are so similarly shaped? To answer that question, we need to go all the way back in time to Euclid, the great-great-great-grandfather of all geometers – the mathematicians who specialise in shapes.

Euclid lived in the ancient Greeks' golden age, when the upper echelons of society could realise their dreams of eschewing manual labour so that they could live a life of leisure and philosophical pursuits. As such he embodied one of the Greeks' most precious beliefs: that everything they saw in the world around them was an imperfect reflection of an idealised version that existed in some divine realm, separated from mere mortals and only experienced through the life of the mind. Every tree was a corrupted copy of a perfect tree. Every human-made building was a dim simulacrum of some divine edifice inhabited by the Olympians.

Euclid extended this idea to shapes. For instance, humans have been making wheels for centuries. But even the manufacturing techniques of today – let alone those of ancient Greece – are not capable of making perfectly circular wheels. They are close enough to circular for our purposes – they will spin on an axle and move a car just fine – but if you take a magnifying glass to a wheel, you will find it littered with dents and bumps that spoil its curved edge. Euclid reasoned that there was a perfect circle somewhere out there – and though he couldn't see or touch it, he could study and understand it using a few rudimentary drawing tools, namely the straight edge and a pair of compasses. In fact, Euclid's obsession with these shapes (and his rigid insistence on the equipment that could be used to produce them) is one of the primary reasons why, centuries later, a pair of compasses – useful for little else than annoying your friends by prodding them with the sharp points – is an essential part of every student's geometry set.

Perfect circles with infinitely round edges; perfect triangles with totally straight sides; perfect squares with inscrutable right angles in every corner. Euclid loved these shapes because they followed such simple rules and effortlessly created such elegant patterns. They are around us all the time, and in fact one of the most common places you'll find them is right beneath your feet. Whenever you're out and about, keep your eyes open for interesting tiling patterns on sidewalks and pavements – you might be surprised at how beautifully designed some of them are!

But in Euclid's world, even shapes without the regularity and structure of those tiles seem to exhibit almost magical properties. For example, grab a sheet of paper and place four dots at random, anywhere on the page. Now connect the dots with a ruler and you've created a four-sided polygon, also known as a quadrilateral.

Even if your quadrilateral looks fairly unremarkable, it is hiding a secret pattern. Use your ruler again to find the midpoint of each side – that is, the point that is exactly in the middle. Since you have four sides, you'll have four midpoints. Now join them up. What have you created? The shape you're looking at in the middle of your irregular quadrilateral is a perfect parallelogram! The sides that are opposite to each other are both exactly equal – that is, they're the same length (go ahead and measure them!). They are also precisely parallel – pointed in exactly the same direction (you can extend them forever in both directions and they will never meet).

Go ahead and try again. Draw another quadrilateral on a new page – try as hard as you can to make it look weird and unusual – and if you join the midpoints of each side, you can't help but form a flawless parallelogram. No wonder Euclid was so transfixed by these shapes – if this is what a shape with no discernible structure or special features can do, imagine what some of the more specialised figures are capable of.

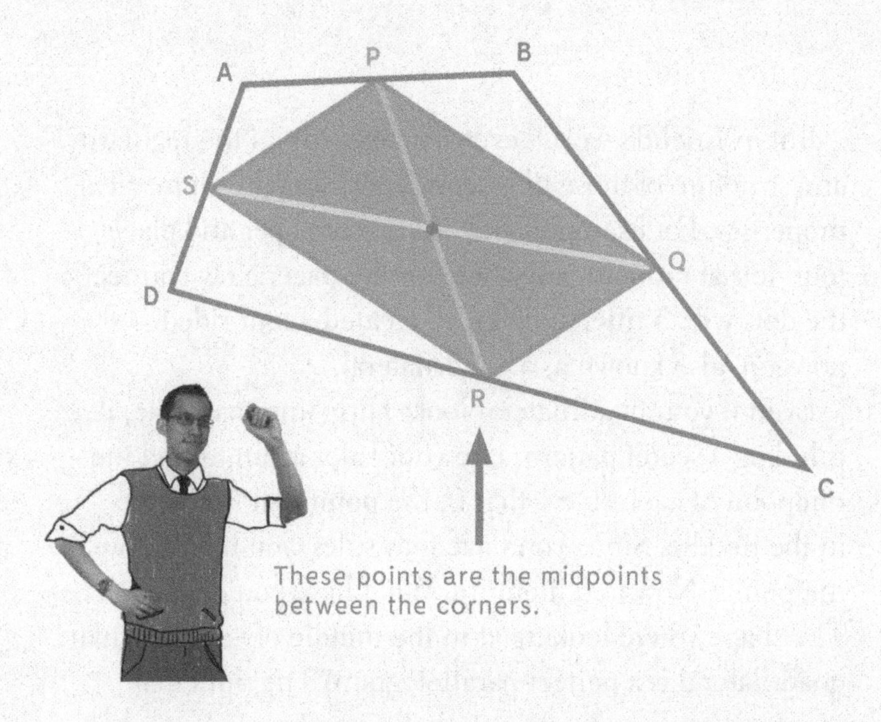

Though I've been making a big deal about points and midpoints, the truly defining feature of Euclidean geometry is smoothness. If you zoom in on any side of a Euclidean polygon, you'll find a flawlessly straight line. If you took a magnifying glass to the face of a Euclidean solid, it would look like a surface that had been ironed over a thousand times. Even curved figures seem to give way to flat smoothness if you zoom in far enough. For instance, if we look closer and closer at the edge of a circle, its curved arcs seem to look like straight lines on small scales:

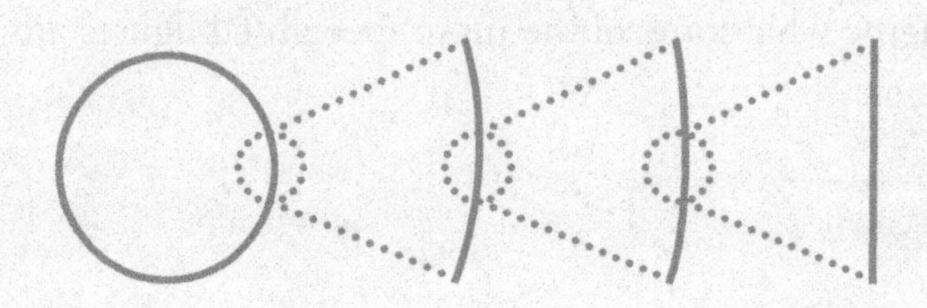

This shouldn't surprise us that much, since we live on an object that is basically a giant sphere that is incredibly flat on the scale that we're familiar with.

Human engineers even use this trick in reverse, constructing objects entirely from straight components – like rigid steel beams – but which look, from a distance, as if they are curved.

Sydney Harbour Bridge

But Benoit Mandelbrot – a Polish mathematician and geometer who did most of his work in the 1950s and 60s – found this Euclidean vision of the world deeply unsettling. Reality is not made up of smooth, unbroken lines. It is filled with jagged edges and bumpy surfaces, lines that have been broken and split into millions of different pieces and have little resemblance to Euclid's divine shapes. A striking example of this gave birth to a mathematical puzzle known as the coastline paradox. It starts with a simple question: how long is the coastline of Australia?

To see why this question leads to a paradox, have a look at the map above. You'll notice immediately that the border between land and sea seems to violate Euclid's rules for geometry in a particularly violent way. There isn't a single straight line (or even a nicely curved line) in sight. You couldn't draw this map with a straight edge or pair of compasses if you had a million years to trace out all the individual sections. The reason this is a problem is because it presents a challenge for us when we try to measure its length.

Imagine you had a gigantic ruler that you could stretch out over the country in order to measure the coastline. Obviously, rulers are straight, so it will miss out some of the crags and outcroppings, but it should still give us a useful approximation of the distance around the edges of the country. Here's what it might look like if we had a ruler that was 1000 km long.

If we wanted to increase the accuracy of this measurement, we would simply use a smaller ruler. If we had a ruler that was 500 km long instead, here's how things would look.

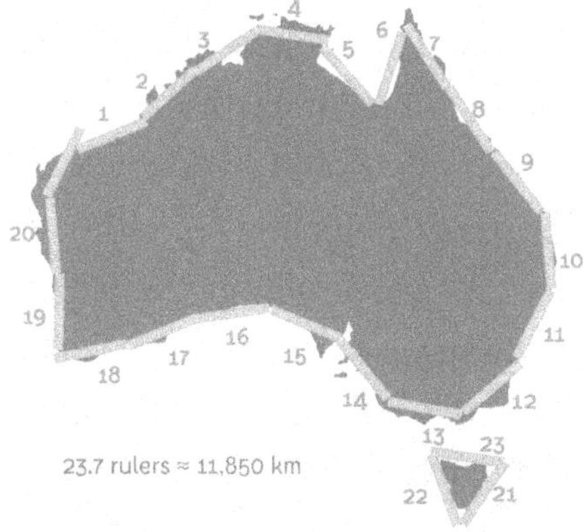

No problem – our new estimate is longer. That's not especially surprising since, as we mentioned earlier, we're now including lots of smaller sections that we had ignored previously because our ruler was too large. For instance, now that we have an appropriately sized ruler, we can finally include Tasmania in the Australian coastline! But that begs the question – what happens if we keep going? What happens if we use an even smaller ruler, say one which is 100 km in length?

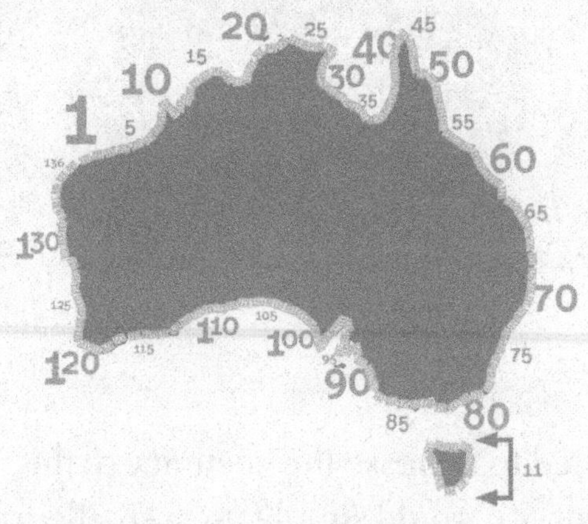

147 rulers ≈ 14,700 km

Okay – this is starting to become a bit worrisome. While we were probably expecting a longer estimate than our previous result, it was also natural to hope that we would start to approach a number that would represent a reasonable place to stop in terms of calculating our estimate of the Australian coastline. This is something we will observe in the next chapter, where we explore rates of exponential growth, which seemed to have a fixed 'cap' (the number *e*). But on the contrary, when it comes to coastlines we find that the estimates keep rising, with no sign of slowing down. Indeed, if you continue to shrink the ruler, the coastline's 'length' continues to increase with no limit. If you had an infinitesimally short ruler, you would find the coastline to be . . . infinite. This is the coastline paradox.

The reason why coastlines do this brings us back to our friend Benoit. He noticed that while a coastline may appear to be a jagged mess, it was clear to him that there must be some rhyme and reason to the madness, even if it was not immediately obvious – just like with our earlier quadrilateral that seemed to magically produce parallelograms. For instance, have a look at this beautiful section of coastline from the Mergui Archipelago in Myanmar:

An aerial photograph like this is quite arresting. And for Mandelbrot, it was objects like these that posed an interesting problem. **Is this geometry?** It certainly doesn't match any of Euclid's ideas: there are no straight lines, no well-defined angles or polygons. But at the same time, there is no randomness here. There are well-defined structures and geometric patterns that are undeniably there, even if they can't be described in a strictly Euclidean way. So Mandelbrot searched for a way to understand and express the shapes he was seeing. He was certain that this was a kind of geometry, but it was not like anything he had encountered before. Since it was filled with shapes that seemed to be broken into many pieces – fractured – he called these objects 'fractals'.

The coastline paradox arises from the fact that coastlines are a type of fractal. Let's go back to that map of Australia, but this time instead of measuring, let's just zoom in to try and understand the shape of the coastline.

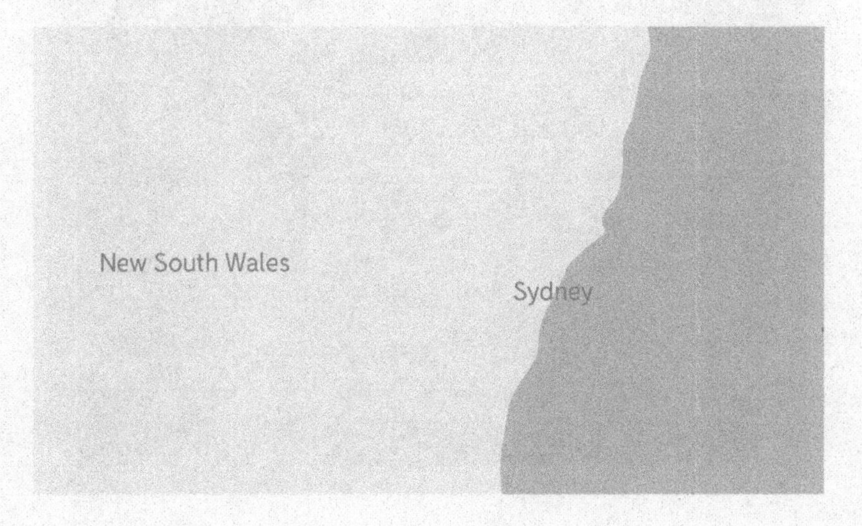

The coastline of New South Wales, like the coastline of Australia as a whole, is bumpy and jagged. And if we continue magnifying, say to the coastline of Sydney, we can observe that the pattern continues:

It doesn't matter how far we zoom in. Each time we magnify the image, we will find new nooks and crannies that make the coastline appear 'fractured'. Even if we were to head down to one of Sydney's beaches and take a photo of the ground from eye level, we would find rocks with holes and outcroppings that would mirror the bumpiness of the first map we looked at. Mandelbrot's key insight was that this is what gave the shapes of nature their characteristic appearance: objects appear similar to themselves on multiple levels of magnification. While Euclidean geometry is characterised by smoothness, fractal geometry is characterised by what mathematicians call self-similarity. Objects resemble themselves even when you zoom in on them repeatedly. And once you recognise this idea, you start to see it everywhere.

Fractals truly are the geometry of nature.

Nature's geometry

And fractals are the reason that your blood vessels look like lightning bolts. You see, while they are doing it for incredibly different reasons and through totally different mechanisms, blood vessels and lightning bolts are actually all about solving the same kind of mathematical problem: distribution.

Blood vessels are doing this by design – their job is to deliver oxygen and nutrients to every tissue in the body (as well as to take away waste products), and so they have been bent by evolution into the most efficient pattern possible for reaching every cubic centimetre of muscle and organ. This pattern must be self-similar because it has to seamlessly expand with the growth of the body depending on it for survival, without fundamentally changing its shape or structure. Self-similarity is also invaluable to our vascular system because it enables arteries and veins to traverse significant distances within our bodies – an average adult will have over 150,000 kilometres of blood vessels snaking through their tissues – while also having components small enough to serve tiny clusters of individual cells, far too small for the naked eye to see. This is why, under the microscope, we can see that blood vessels branch out into a seemingly innumerable set of offshoots and tributaries – because they need to supply blood to every cell in the body.

Once you put it this way, it becomes less surprising that blood vessels should be fractal. But what about lightning? No one is designing the lightning into one shape or another – are they? In fact, the answer to that question doesn't even matter – because you can understand the fractal DNA in a lightning bolt just by thinking about what a bolt of lightning actually is. Electrical charge builds up in a storm cloud from the constant friction of water droplets swirling around inside it, just like you can build up charge in yourself by rubbing your feet back and forth against carpet. When that charge exceeds the cloud's ability to hold on to it, it erupts from the cloud like lava from a volcano and follows the easiest path it can find to reach the ground, where it can dissipate.

But as it makes its way to the earth, the lightning bolt finds small pockets of air molecules that have just enough negative charge to attract the flow of electricity a little more than the air around them. This causes the lightning bolt to twist and bend, losing portions of its energy and becoming narrower as it does so. Sometimes there are two or more air pockets that are equally attractive to the lightning, so the bolt dutifully splits off into sections and heads towards both, again becoming thinner in the process. Since the same physical laws govern large lightning bolts and small ones, the same geometric moves are made by the mother bolt as by its daughters as they split off, giving it an unmistakable self-similarity. The lightning bolt's reason for existence is to distribute that tremendous electrical energy – just like your blood vessels' purpose in your body is to distribute life-sustaining blood – so it cannot help but adopt a branching fractal structure to bring about its own demise, the very same structure your body has adopted to keep itself alive.

CHAPTER 05

TEACUPS AND (ALMOST) INFINITE MONEY

It's still dark outside, and the house is blissfully quiet. My children are all still in their beds – for now – so I creep downstairs and slowly fill up my kettle. I hit the switch. As the water temperature rises, I can hear the bubbles jostling around and bursting inside the kettle.

The switch clicks, and I waste no time. The kettle leaps off its stand and I immediately pour the scalding water into my waiting cup with the teabag and sugar already in place. Why the hurry?

The boiled water in my kettle – like all objects hotter than their environment – cools down in relation to its temperature. The hotter it is, the faster its temperature drops. Here's what happens when I measure the water with a thermometer, starting from the moment it comes off the boil. In the first 60 seconds, it drops by a staggering 35 degrees. Compare that to what happens after the water has been left to stand for ten minutes; in the next 60 seconds, it becomes only 3 degrees cooler. What's going on?

It's actually not the object's temperature that matters, but rather the difference between the object's temperature and its environment. The greater that difference, the faster the difference changes – in both directions. This is the same for thawing as it is for cooling. When it comes to temperature, all objects in the universe cave to the most brutal peer pressure: **every object wants to become the same temperature as the objects around it.**

So, water cools faster when it's hotter. But my cup of tea isn't the only object in the universe to behave like this. It's the same for things that grow when they have no external limitations. For instance, take a virus that has just made landfall on a poor and unsuspecting host. In its first few hours of existence, before the immune system takes notice and mounts a counterattack, the virus hijacks a few of the host's cells to manufacture copies of itself. But once those cells erupt, all hell breaks loose and millions of intruders flood into the bloodstream, striking far more cells than in the initial strike. Each of these new cells becomes a new engine for conquest and numbers of the virus increase even faster. In other words, viral infections grow more quickly the larger they get. (Obviously, up until a point – or so you would hope!) This characteristic is called exponential growth (or in the case of my cup of tea, exponential decay).

Or consider money that's been left in a term deposit. During primary school, I remember my local bank encouraging us to open accounts by enticing us with free stickers and other trinkets. But all I had to deposit was a few measly dollars. I still recall the first time I received my bank statement with excitement, only to discover that in my first month I had earned the princely sum of . . . 3 cents in interest. Yet despite this initially underwhelming return, the law of exponential growth means that faster growth comes from a higher bank balance. The more I've saved, the more I'll earn.

This isn't surprising to most people – it's the basis on which our entire economy is built, after all. But exponential growth harbours a dark secret. You see, while exponential growth seems to hold the promise of unencumbered financial success, it actually has a limit. Let me explain.

Suppose you have $1 to put into a bank account, and through incredible luck you find a bank that offers to pay you 100% interest per annum. 'They must be taking pity on me for my pathetic savings,' you reason to yourself. Sitting in the bank manager's office, you start to do some calculations out loud. 'With interest of 100% per annum I'll earn a whole extra dollar in interest, leaving me with $2 by the end of the year,' you conclude.

Compounding period	How many of those periods are in a year?	How much interest do you get paid each time?	What will your account be worth at the end of the year?
Year	1	100%	$2

The manager smiles at you. 'It could be even better than that,' she says. 'You'll end up with $2 if your compounding period is the whole year. But we'll let you choose any compounding period you like!'

What difference does it make? Then suddenly you realise: choosing a shorter compounding period can help your account to grow faster, even with the same original investment and an identical interest rate. How can that be? Well, having shorter compounding periods will mean that you'll earn less interest each time, but you'll be paid interest more frequently. Crucially, each interest payment will be based on that ever-increasing amount – so it increases gradually in size with every compounding period. Here's what will happen if the compounding period is six months instead of a whole year:

Compounding period	How many of those periods are in a year?
6 months	2
How much interest do you get paid each time?	**What will your account be worth at the end of the year?**
50%	$2.25

Not a bad profit! But if this is the benefit you get when compounding twice a year, what if you compounded more frequently? Say, every month?

Compounding period	How many of those periods are in a year?
Month	12
How much interest do you get paid each time?	**What will your account be worth at the end of the year?**
8.33%	$2.613035 . . . (rounds to $2.61)

Still an improvement on the previous result; but now that you've had a taste, you want more. Why not compound every single day?

Compounding period	How many of those periods are in a year?
Day	365
How much interest do you get paid each time?	**What will your account be worth at the end of the year?**
0.27%	\$2.714567 . . . (rounds to \$2.71)

It's worth pointing out that there are some fairly counterintuitive things going on here. For starters, the interest rate looks laughably tiny on the situation we're considering here. 0.27% seems so small that it feels like it is hardly worth mentioning. For example, I'm 178 cm tall, so 0.27% of my height is less than half a centimetre. If two people were to stand beside each other and there was only half a centimetre of difference between their heights, I would not be surprised if you simply said that they were the same height. So 0.27% seems negligible. But – and this is one of the central mathematical tenets of compound interest that makes it so powerful – when it is applied so many times (365 in a single year), even a small change can lead to significant results.

The other thing that defies our intuition is about how much money we are gaining as we make the compounding period smaller. When we changed from once a year to twice a year, we doubled the compounding frequency and gained 25 cents. But when we changed from monthly to daily, we increased the compounding frequency by more than 30 times and gained only 10 cents.

And if you thought that was bad – watch what happens when you try to increase the compounding frequency again.

Compounding period	How many of those periods are in a year?
Minute	525,600
How much interest do you get paid each time?	**What will your account be worth at the end of the year?**
0.00019%	$2.718279 . . . (rounds to $2.72)

If you have the audacity to compound not just every month or every day but every minute, you have increased the compounding frequency by a whopping factor of 1,440 – but in terms of what you've gained, you've only increased your account's value by less than a single cent. (It only goes up because we tend to round currency to the second decimal place.)

What we're experiencing here is similar to something called the law of diminishing returns. We can keep on compounding ever more frequently, but the gains in our bank account are getting smaller and smaller. Since we've come this far, we might as well illustrate this further by going one more step:

Compounding period	How many of those periods are in a year?
Second	31,536,000
How much interest do you get paid each time?	**What will your account be worth at the end of the year?**
0.0000032%	$2.71828178 . . . (rounds to $2.72)

That number in the bottom right is the one I'm most interested in. Let me summarise what we've seen so far in this financial example by just looking at that:

Compounding period	What will your account be worth at the end of the year?
Year	$2
6 months	$2.25
Month	$2.613035 . . .
Day	$2.714567 . . .
Minute	$2.718279 . . .
Second	$2.71828178 . . .

Have a look at the final value of the bank account. Do you see how it rises from $2, but starts to converge towards a particular value rather than increase forever? Mathematicians call this idea a 'limit' – because you can see that the account's value behaves as though it has some kind of constraint acting on it, limiting it from growing any further.

The number that you can see emerging from this limit (2.7182818 . . .) comes up wherever there is exponential growth (like our bank account) or exponential decay (like my cup of tea as it cooled down). This number is so important that it gets a special name: 'e'. You can take that to mean 'exponential', or as an abbreviation of the other name it's been given: 'Euler's number' (referring to the Swiss mathematician, Leonhard Euler).

To me this speaks of a cosmic mystery.

This number, like its more famous cousin pi (π), seems to be baked into the laws of the universe. Just as all life is made up of DNA written in the same four organic molecules – guanine, adenine, thymine and cytosine – all exponential growth and decay in the universe is built from the same DNA, except it is all written in terms of the number e.

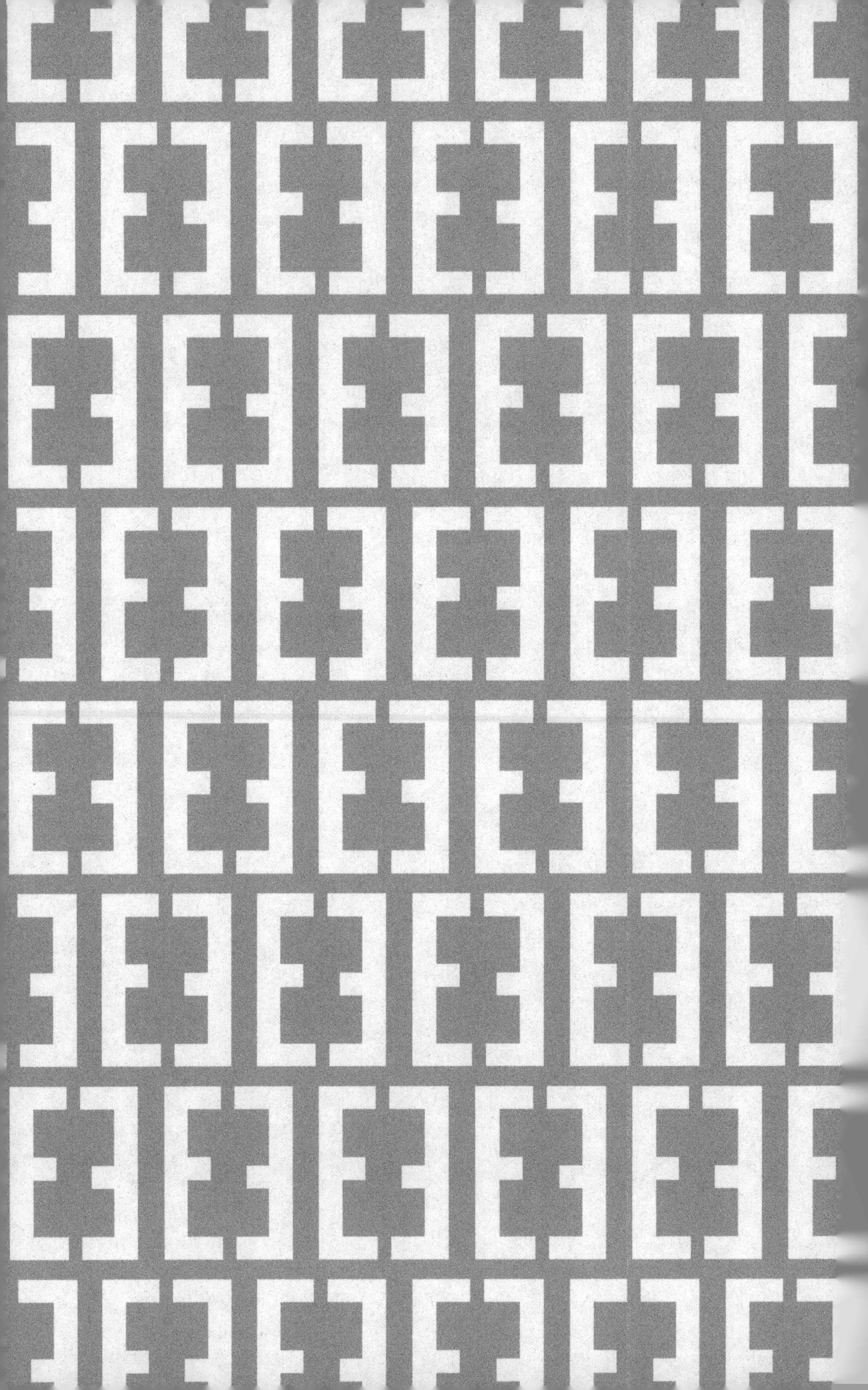

CHAPTER 06

E IS A MAGIC NUMBER

The number *e* is truly underrated. Almost everyone has heard of pi (3.14159265 . . .), and it even gets its own day of celebration (Pi Day is usually marked on 14 March – written as 3–14 in the United States). There are also countless pieces of art dedicated to pi around the world.

Suffice it to say, pi gets plenty of love.

One of the things that all truly wonderful numbers do, though, is come up in unexpected places. And *e* is definitely no slouch in this department. I want to show you one of the most surprising places that *e* appears in everyday life. To help me, I'll need one of the world's most popular foods: chocolate!

Life is like a box of these!

Chocolates are made and sold in many different forms, but I want you to picture a box of chocolates – the kind where you have them all laid out individually in a tray. Imagine opening them up and savouring the smell as you ponder which kind you'll try first.

You're about to select your first piece, but then you remember that you should probably offer to share them with your loved ones before stuffing your face with sugary goodness. So you get up with the open box in your hands, only for disaster to strike – you trip over your own feet and the entire box of chocolates ends up spilled onto the floor.

Did anyone see you? Your eyes dart around, and no one is in sight. What a relief! All you need to do is put all the chocolates back in the box. You don't remember where they were exactly, so you position them randomly in the tray until they look reasonably untouched.

Here comes the question. You know it's virtually impossible for all the chocolates to have been put back in their original spots – that would be incredible luck. It's likely that most of the chocolates are in the wrong spots. But what's the chance that every single chocolate is in the wrong spot? What's the chance that *no* chocolate has ended up where it began?

You might not know where to even start in trying to answer such a question. Here we brandish one of the mathematician's favourite problem-solving techniques: tackle a simpler version of the question first, and see if you can observe a pattern or structure that can help you with the harder original question.

Often the easiest way to make a question simpler is to make it smaller. So let's start with the smallest this question can possibly be: what if there was just a single chocolate in the box?

Let's name each chocolate so we can keep track of where it came from and where it ends up. On the left, you can see how the chocolates began; on the right, we see all the ways they could be rearranged after you drop them and put them back.

With just a single chocolate in our box, we've perhaps simplified the situation a little too far! Not much interesting stuff is happening at all. In fact, since there is only a single spot for our lone chocolate to occupy, it has no choice but to go right back where it came from. This means that if we have one chocolate, there is a 0% chance that it will be in the wrong spot.

Let's liven things up a little. What happens with two chocolates?

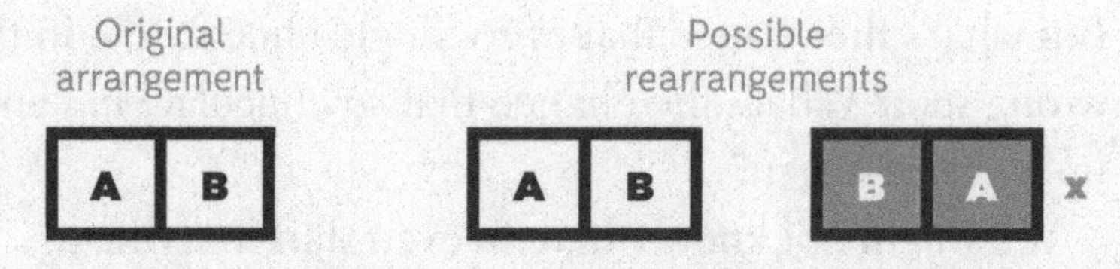

On the right, you can see we now at least have some options. One of the options is always for the chocolates to resume their original positions – but in this case, the only other option is for them to be reversed. This is exactly what we want – every chocolate (all two of them!) in a 'wrong' spot, different from where it started. Out of the two possible rearrangements, one of them meets the condition we want – so there is a 50% chance of getting everything in a wrong spot.

Let's go again! Here's what happens with three chocolates:

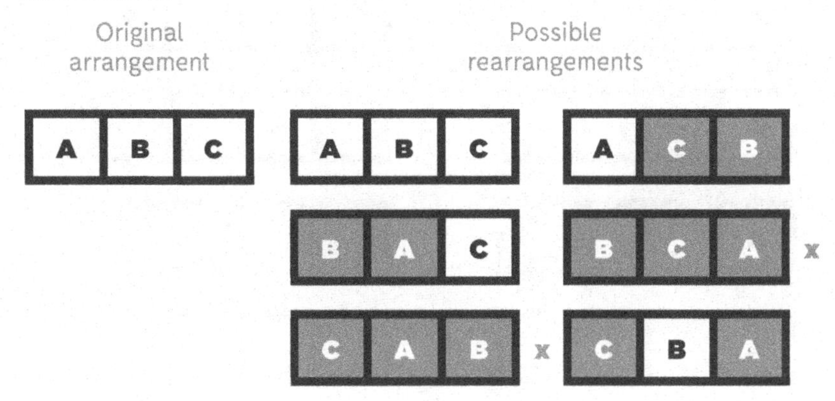

Now things are starting to get really interesting. You can see I've highlighted any chocolates that have been placed into the wrong spots. In some situations, like BAC, you have two chocolates in the wrong spot (B and A) but one chocolate (C) has found its way to its original spot. A careful check shows that out of the six possible rearrangements, two are the ones we are interested in. Two out of six: that's roughly a 33% chance.

Let's really turn the dial up now: what happens with four chocolates?

Original
arrangement

Possible rearrangements

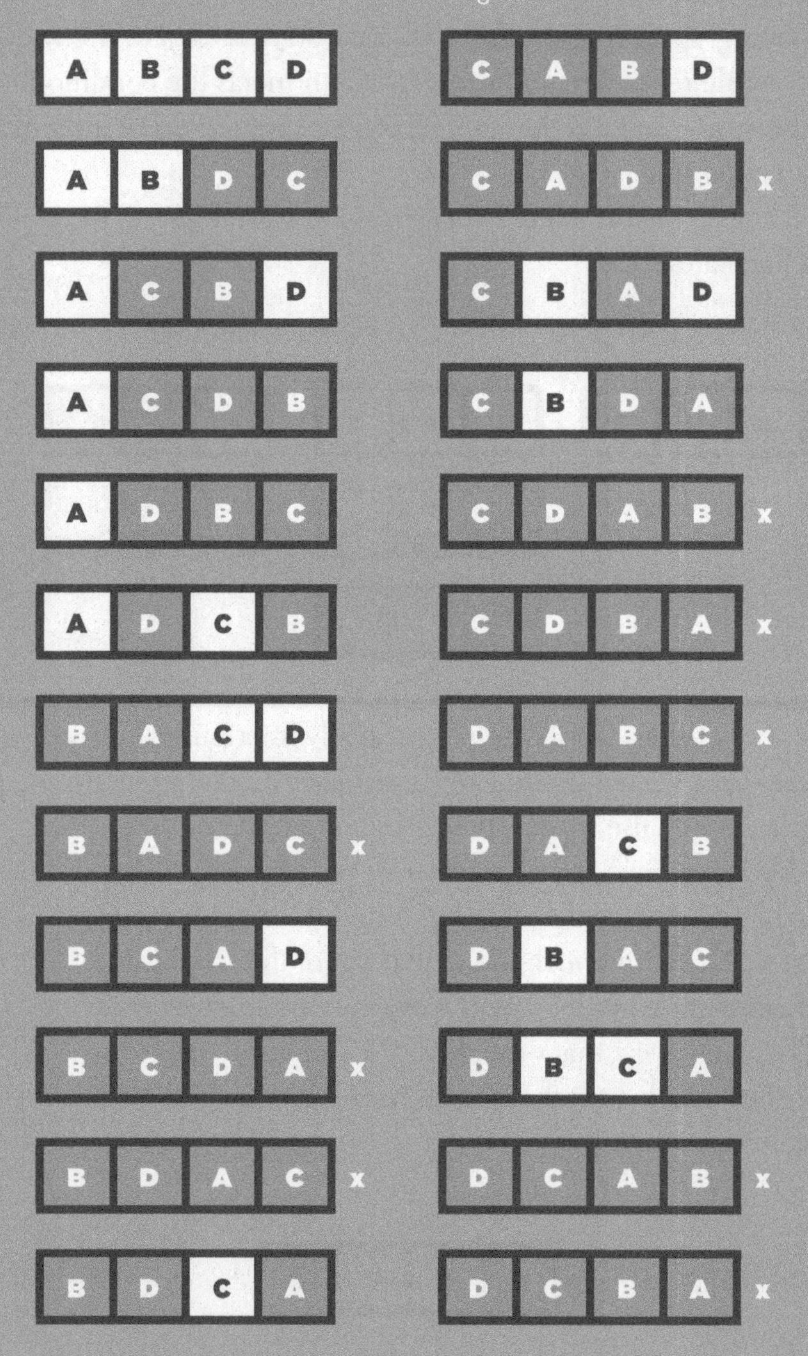

This is starting to get genuinely complicated. There are now 24 different ways that our chocolates can go back in the box. I have gone through and counted, and there are exactly nine of those rearrangements where all four of the chocolates are in the wrong position. Nine out of 24 gives you a chance of 37.5%.

Where is this going? I won't draw all the permutations that emerge if you add another chocolate – there are 120 different ways to rearrange five chocolates! But here are the numbers if you continue the pattern:

Number of chocolates	Number of ways to rearrange them	Rearrangements where they are all 'wrong'	Chance of rearranging into one of the 'all wrong' patterns
4	24	9	37.5%
5	120	44	36.66666 . . . %
6	720	265	36.80555 . . . %
7	5040	1854	36.78571 . . . %
8	40,320	14,833	36.78819 . . . %
9	362,880	133,496	36.78791 . . . %
10	3,628,800	1,334,961	36.78794 . . . %

'Okay . . . so what? You said you'd show me something to do with *e* – I don't see it anywhere in these numbers.'

Right – there's one step left. You can verify it yourself if you have a calculator (the one on your phone will do). First, you need to know that the number *e* is equal to the following:

$$e = 2.718281828459045\ldots$$

If you go ahead and punch '100 ÷ *e*' (where you type the number instead of the letter, although even mobile phone calculators seem to have an *e* button these days!) into your calculator, what you should find is the following:

$$100 \div e = 36.7879441171\ldots$$

Now check the table on the previous page again. Does this number look familiar?

There's some pretty **intense** mathematics underneath why this number, which we met in the context of exponential growth, should appear in these calculations to do with mismatched chocolates. It's a little long-winded for the scope of this book but what I want to highlight for you is the way that mathematics can find a connection between objects that seem so completely different on the outside. Mathematics allows us to see deeper into realities until we can recognise what it is that things truly have in common, just like the way chemistry allows us to recognise that diamonds (like the ones in engagement rings) and graphite (like the stuff inside pencils) are both made of carbon.

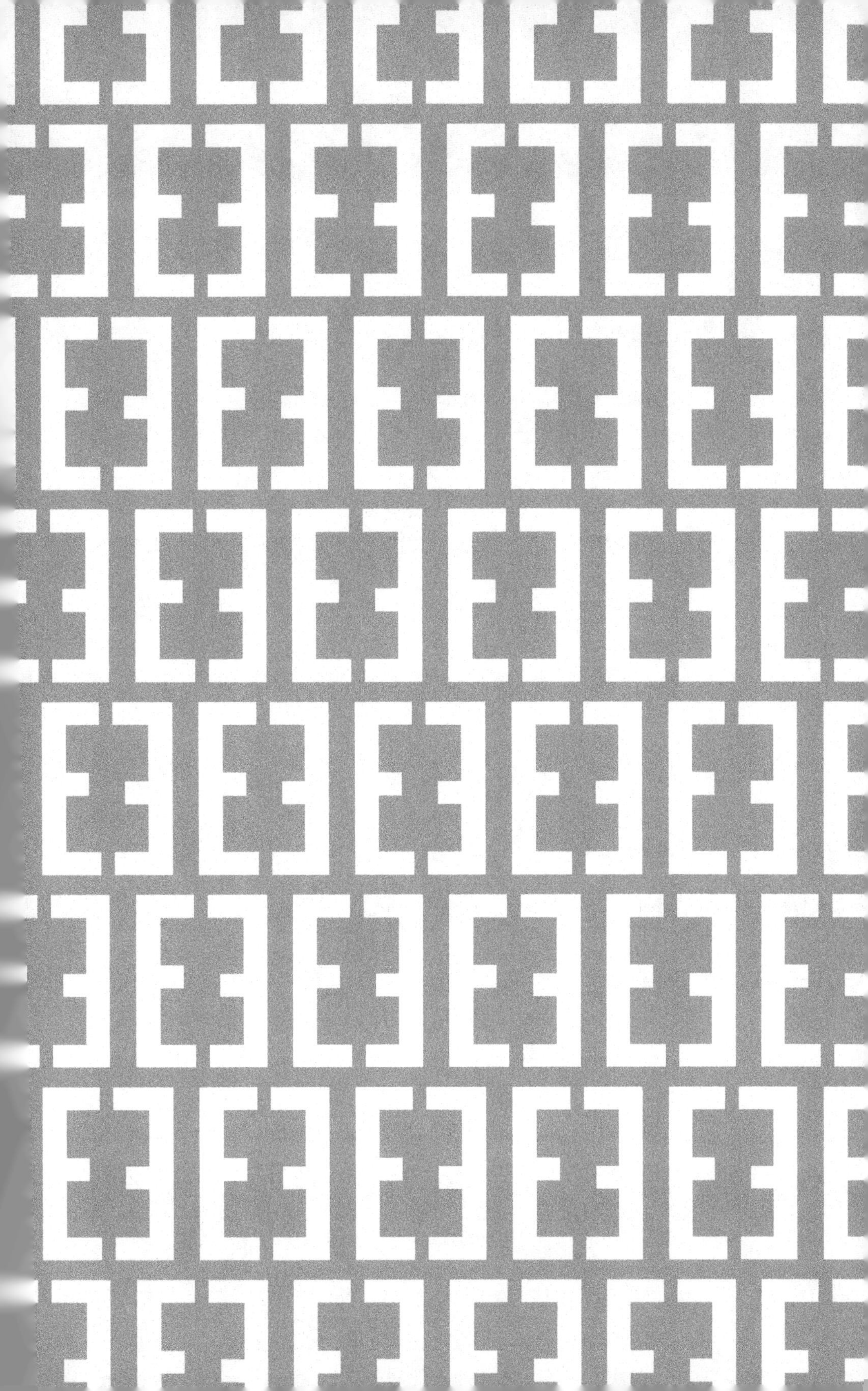

CHAPTER 07

WHAT SUNFLOWERS KNOW ABOUT THE UNIVERSE

There is something mesmerising about watching synchronised swimming. In high school I was on the school water polo team, where I experienced firsthand just how difficult it is to merely remain stationary with your head above water. Synchronised swimmers accomplish something that is immensely more difficult: they perform complicated manoeuvres, many of which involve remaining submerged without the ability to breathe. This is amazing enough on its own, but the truly captivating thing is that they pull off these feats in perfect timing with their partners. Sometimes you can see ten swimmers, all under water, all simultaneously dipping and twirling as if driven by a single mind.

It takes months of gruelling practice for a team of synchronised swimmers to get to the point where they can achieve this. And it requires more than just dedication and discipline to construct and execute a routine like this – deep thought and artistry are needed to create a truly beautiful routine that all the swimmers can perform.

That's why I find it nothing short of astonishing that there are billions of objects around the world that work, each and every day, to create patterns whose beauty rivals or even exceeds those of the synchronised swimmers. It's pretty likely that you've personally walked past thousands of these over the course of your lifetime and never even noticed their dazzling designs unfolding before you each day. They perform their routine with a unity that would make even the greatest Olympic swimming teams shudder with envy, yet they do so without conducting a single practice. In fact, they don't even really communicate with one another in

Human symmetry

any substantial way to coordinate their efforts. Who are these effortless masters of synchrony? I'm talking about sunflowers.

'What? Sunflowers don't have limbs, and I doubt they'd be any use if you threw them in a pool! How on earth can they even be compared to synchronised swimmers?' I hear you say. By now, hopefully you're not too surprised by the answer: mathematics.

Nature's symmetry

Look closely at the pattern of florets covering the centre of the sunflower. Have you ever paid attention to one? Have you ever noticed the amazing symmetry produced so flawlessly by the sunflower? Each sunflower does it, performing its routine with perfect precision throughout its lifetime – just like its brothers and sisters, without ever talking to them. How do they do it – and, just as importantly, why?

To understand what's going on underneath the surface here, we need to first cover some basic horticulture. When we humans build something, we do it in what might be called a linear way. For example, imagine building a brick wall. We start from the bottom, move from left to right, and end at the top. However, like every other living thing on the planet, sunflowers grow organically – that is to say, they start small and grow outwards from the centre.

This one fact is crucial because it is our first step to understanding the logic of the sunflower. Imagine with me, if you will, a sunflower growing in super-slow motion: one floret at a time. What would it look like?

Florets start off in the centre of the flower and get pushed outwards by new florets as they form – that's why the ones closest to the edges of the flower are the largest (because they've been alive and growing for longer). The arrangement of florets will depend on the direction in which each new floret pushes the previous one that was in the centre.

Sunflowers perform their routine with perfect precision.

25% rotation

For instance, if the flower does a quarter turn – that is, 25% of a full rotation – before each new floret grows, then we get the pattern pictured above.

Notice, by the way, how the florets grow. I've numbered them to help you keep track. The oldest florets (which have smaller numbers, because they came first) are the largest and furthest away from

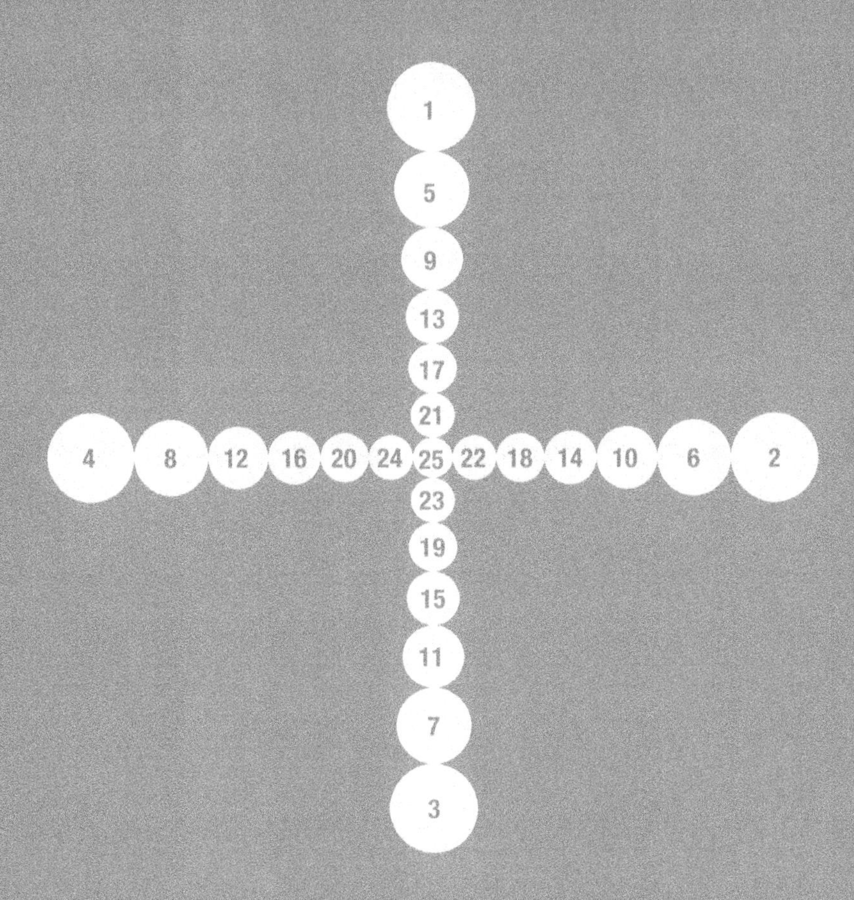

the centre. That's because they've had the longest time to grow (making them larger) and they've also been displaced the most (making them further away).

This is nice and all, but it's a huge waste of space – look at all those gaps where florets could be stored, but the flower is leaving it empty. How about if we do a 20% rotation before we grow the next floret?

20% rotation

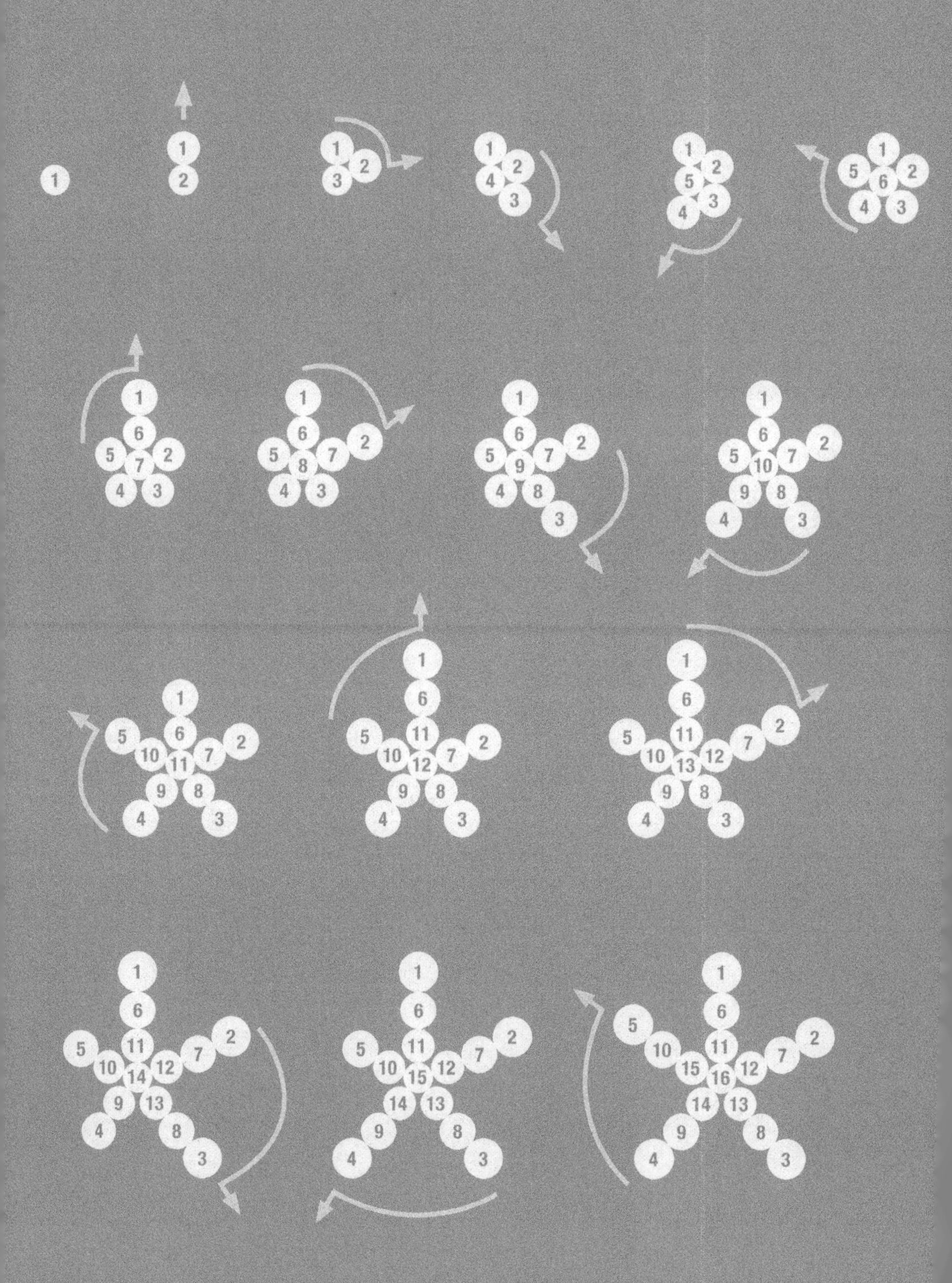

This is better, but still colossally wasteful. How can we improve this design? Well, instead of a 25% or 20% rotation, how about trying a 34% rotation?

34% rotation

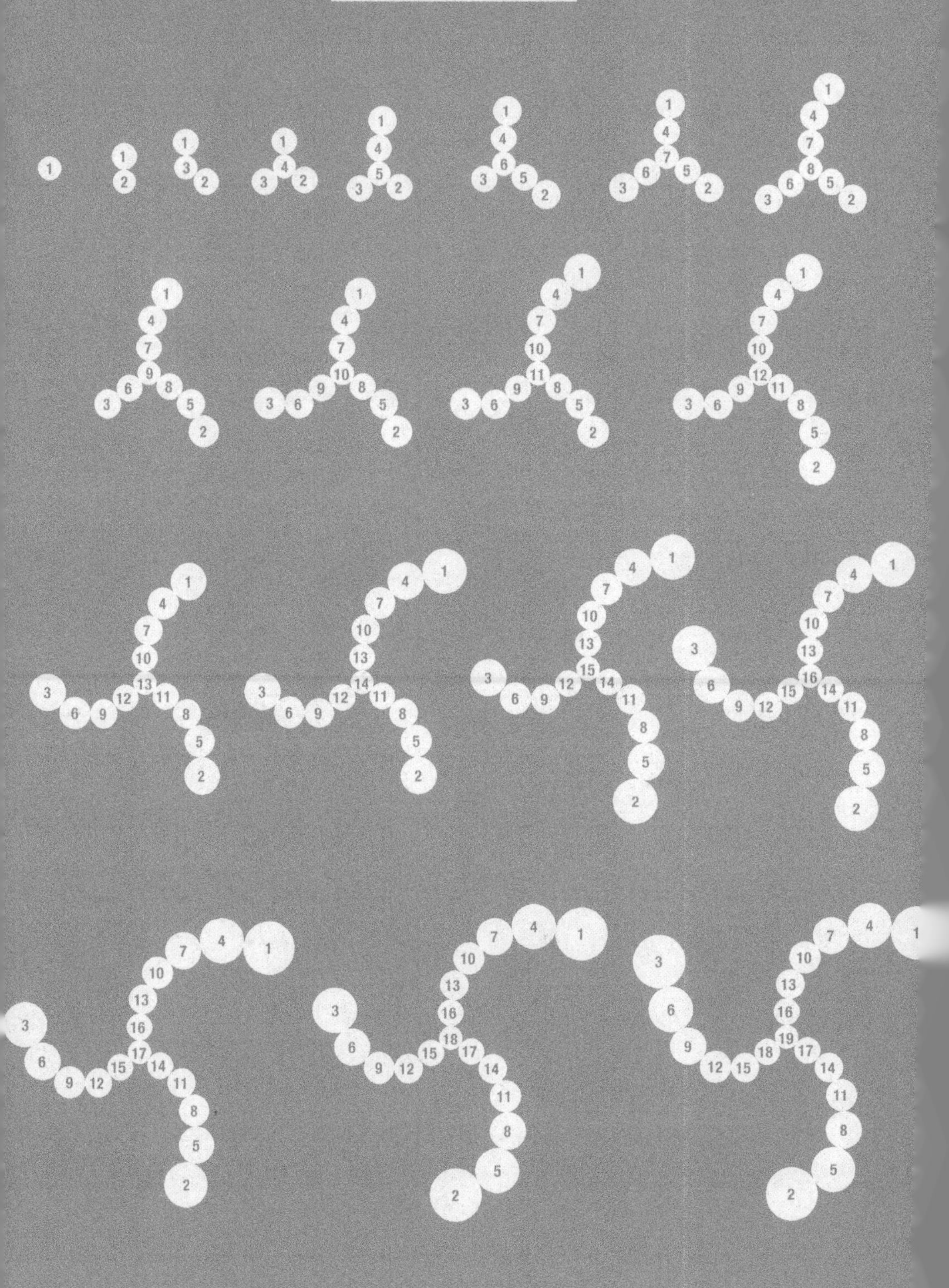

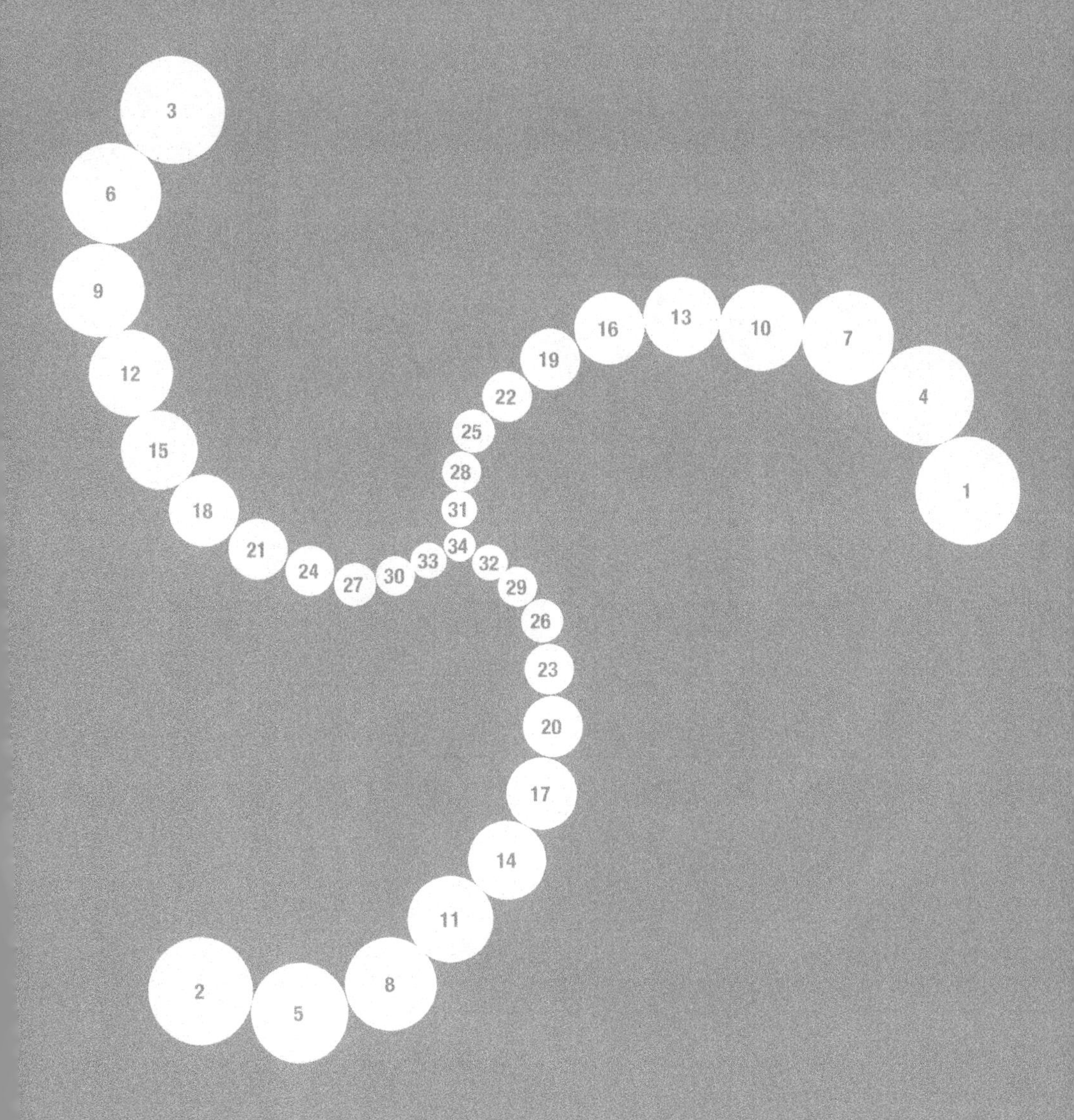

This one is a little more interesting (and flower-like!). Unlike the previous arrangements, which produced straight lines of florets, this one produces curves. The reason why is that each new floret grows after just a little bit more than a third of a rotation – meaning that each time, the newest floret is sent off in a slightly different direction. As a result, the 'rows' curve around rather than line up straight.

17% rotation

We can take this spiralling idea and apply it even more strongly – try this one:

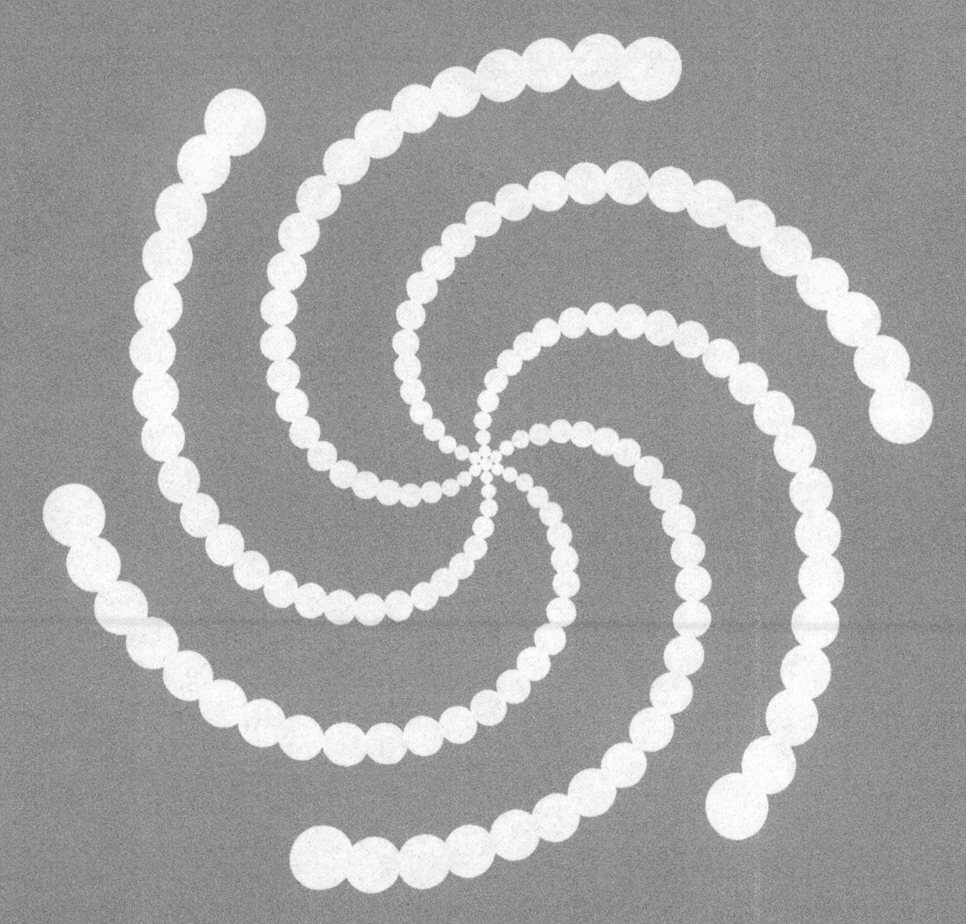

This design comes from a 17% rotation. Since it rotates twice as often as the 34% version, it has twice as many 'arms'. It's a vast improvement on our original design because it can pack many more florets into the same amount of space – good news for the resource-hungry plant, as well as for pollinating bees which have a better chance of seeing it. Is there an ideal rotation to go through that will maximise the number of florets that will fit on the flower?

Answer: yes. Oh yes. This is one of the brightest gems in all of mathematics. I'm about to introduce you to one of the most wonderful numbers in all the universe. It often goes by the unassuming name 'phi' (as in the Greek letter). Its symbol is ϕ, which is thankfully very different from its much more famous and similar-sounding cousin, π (pi). But it's also known by a much grander sounding name:

the Golden Ratio.

As with many other important mathematical objects, the Golden Ratio was studied at length by the geometry-obsessed Greeks. To understand the Golden Ratio, it will help to think of a simple little problem to do with a piece of string.

Picture a string stretched tight between two pins and cut off at the ends. Geometers (the mathematicians who focus on geometry) call this an interval. Now imagine being able to join another interval onto your original, making a new interval that's longer than the original.

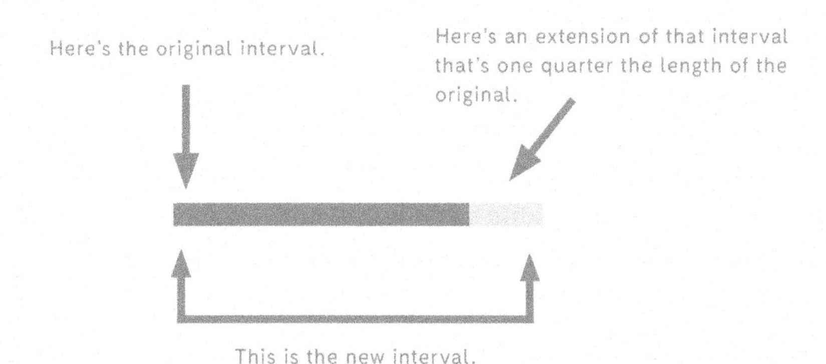

The Greeks wanted to understand: what's the relationship between the original interval, the extension, and the new interval? How much bigger is one than the other? In the diagram on the previous page, the original interval is 4 times the length of the extension and the new line is 1.25 times the length of the original.

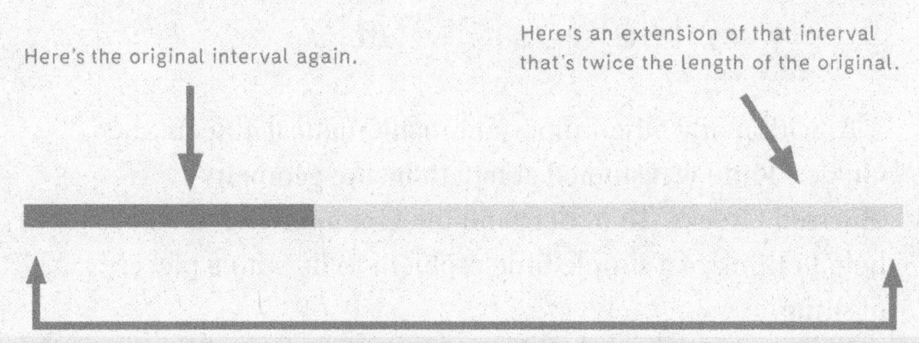

This time, the original interval is half the length of the extension and the new interval is 3 times the length of the original.

The Greeks wondered: how long must you extend the interval so that the ratio between the original line and its extended version is the same as the ratio between the original line and the extension itself? Here, let me draw it for you:

Let's define the original interval
to have a length of 1.

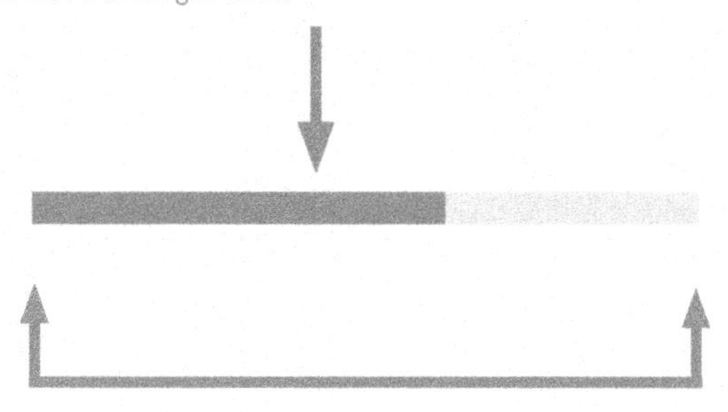

Let's call the length of the new interval Φ
(that's the Greek letter 'phi'). It follows that the
extension must equal $\Phi - 1$ (it's the difference
between the longer interval and the shorter
interval, which are Φ and 1 respectively).

The value of phi is exactly equal to $(1 + \sqrt{5}) / 2$, or approximately 1.6180339887. The digits after the decimal place keep going on forever and never repeat – that is why even 10 decimal places (which would be an insane level of precision in most contexts) is still an approximation.

This simple number gives birth to a variety of shapes that the Greeks recognised as innately beautiful. For instance, if you draw a rectangle whose sides are in the Golden Ratio, then you create a golden rectangle:

The golden rectangle is universally recognised as an aesthetically proportioned figure. This is so much the case that it's highly likely you have a collection of golden rectangles on you right now. If you have a bank card, or perhaps a driver's licence, take it out and put it on the table. If you can find a ruler nearby, go ahead and measure the length of each side. Then, use your calculator (the one on your phone will do) to divide the longer side by the shorter side. Cards all differ slightly in their size, but did you get something close to 1.618?

One of the neatest features about the golden rectangle is that it demonstrates how phi can endlessly create itself. To show you what I mean, consider what happens if I draw a line to divide the golden rectangle into a square and a new, smaller rectangle.

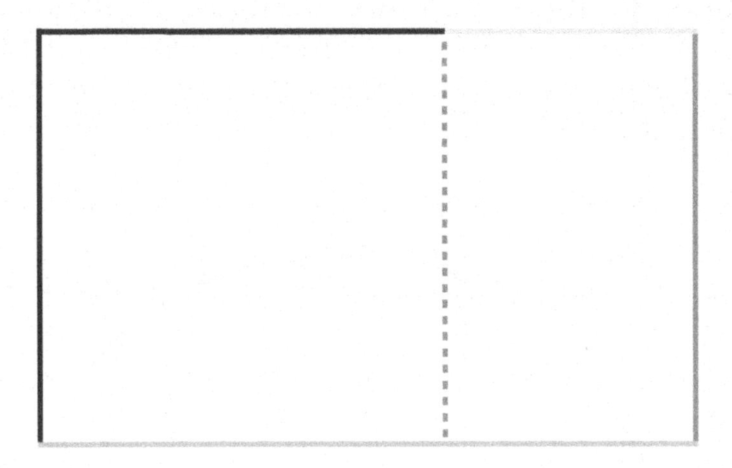

Look closely at the smaller rectangle we've created above on the right. Does it look familiar? It's another golden rectangle! And in fact, we can do this forever – we can make golden rectangles within golden rectangles into infinity. If you repeat this process forever, you'll create a shape we encountered in the chapter 'Lightning through your veins': a fractal!

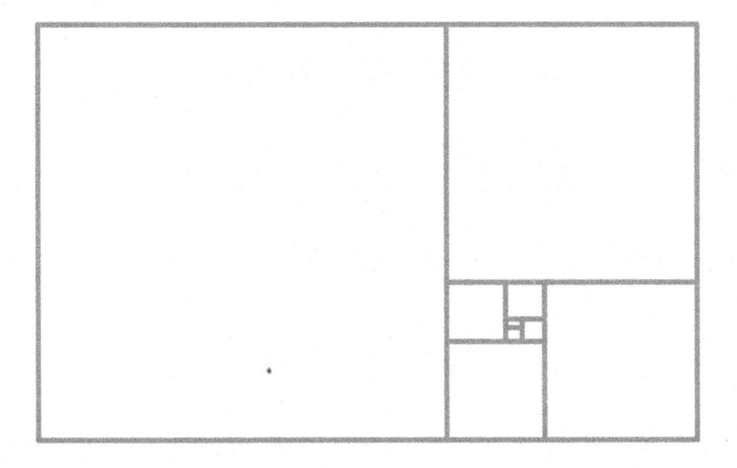

If you join the corners of the squares with circular arcs that fit snugly into the squares, you create an even more striking shape:

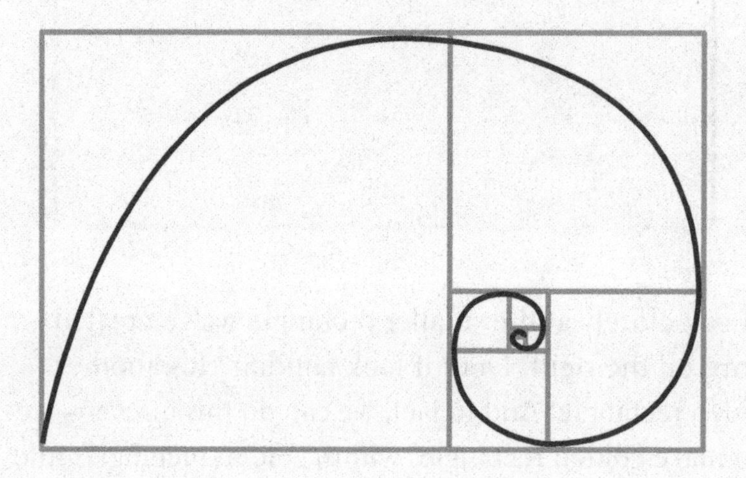

In fact, you probably recognise this one from nature too!

The Golden Ratio can be found littered throughout both nature and human designs. But my favourite by far is what we were looking at to begin this chapter: the sunflower.

To help us see the connection between sunflowers and the Golden Ratio, it's useful to learn something about the way that percentages (like 25% and 17%, the rotations we were playing with before) relate to fractions (like ½ and ¾) and decimal expansions (like 0.83 and 3.14). Fractions, decimals and percentages are just different ways of dressing up the same numbers, and, like us, numbers dress up differently depending on the activity that they are engaged in. Comparing two different quantities? Most likely you'll dress up the numbers as percentages. Want to divide something into shares? This time fractions are probably the most useful. Trying to measure things (like weights or lengths) on a physical scale? Decimals would be the most natural choice.

'Per cent' is literally Latin for 'out of a hundred'. We still use the prefix 'cent' to indicate a hundred – such as when we talk about one hundred years (a century) or a one-hundredth anniversary (centenary). This 'out of a hundred' idea is also where the percentage symbol itself comes from:

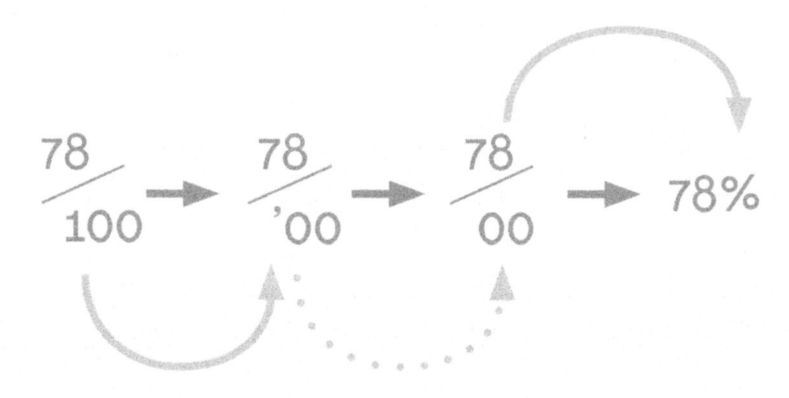

This means that 78% is the same as 78/100, which in turn can be written as 0.78. Similarly, 100% could be represented as 100/100, which is equal to 1. That's because any number divided by itself is equal to 1 – the only exception being zero, since nothing can be divided by zero (if you are curious about why division by zero is impossible, skip forward to chapter 24, 'Math error'). In fact, we can write absolutely any decimal number as a percentage – even numbers bigger than 1. The Golden Ratio, which is approximately equal to 1.618, is a perfect example: it would be 161.8%.

If we choose the Golden Ratio as our rotation – using 161.8% of a full rotation instead of 25% or 20% or 34% – look at the shape that emerges:

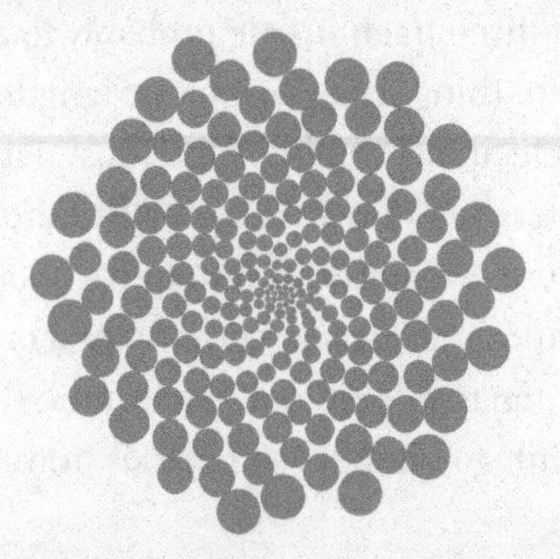

When I saw this for the first time, it blew my mind into a million little pieces. Who knew that the humble sunflower knew about this wonderful mathematical constant which we find written across the face of the universe? And, just in case you want to marvel at this a little more deeply, here's another version of the diagram that makes it a bit easier to see the individual spirals dancing together to make this amazing figure:

A BRIEF NOTE: It's not really that sunflowers are intelligent beings who worked out the equations to come to this conclusion – but simply that any ancestral sunflowers that didn't use the Golden Ratio produced fewer seeds per flower, and were therefore weeded out by the process of natural selection. But the fact that this natural algorithm should arrive at a mathematical truth like this is almost as beautiful as the human beings who calculated it through their ingenuity and insight!

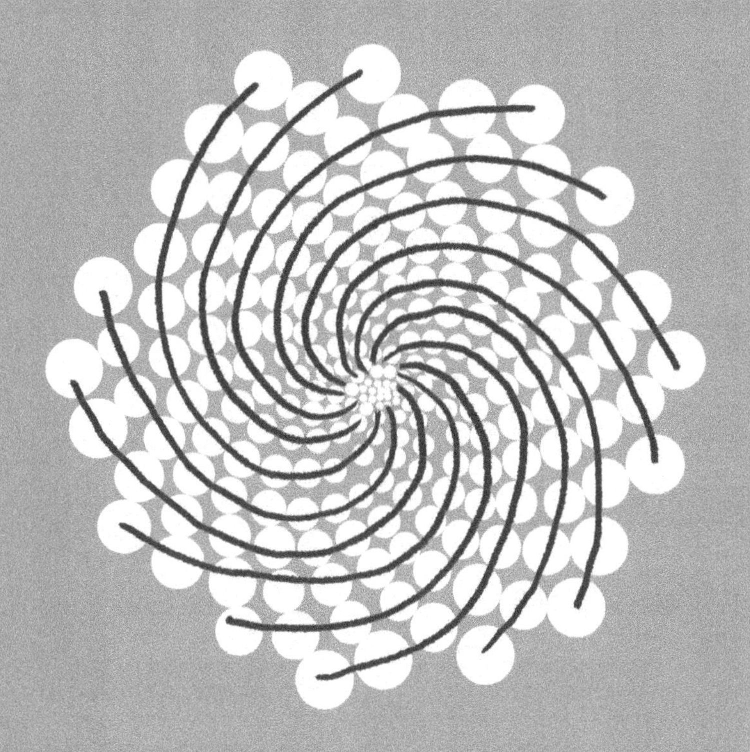

The spirals of florets are packed together with masterful precision. It's been said that if you go down deep enough into anything, you will find mathematics. I think the humble sunflower may be one of the clearest examples of this in all of nature.

CHAPTER 08

WILL THE REAL GOLDEN SEQUENCE PLEASE STAND UP

In the chapter 'What sunflowers know about the universe', we met the **Golden Ratio**, which is approximately equal to 1.618. It's an astonishing number with even more astonishing powers: it's the foundation of what we regard as beautiful, it can create itself geometrically, and it's able to solve evolutionary problems with an elegance that the most discerning structural engineer would envy. The Golden Ratio has spawned a series of spin-offs, each of which has taken on a life of its own: the golden rectangle and the golden spiral were just two that we had a brief look at.

One of the facts that's been thrown around here and there is that the Golden Ratio has a special relationship with that famous list of numbers, the **Fibonacci sequence**. Here's what the sequence looks like:

0, 1, 1, 2, 3, 5, 8, 13, 21, 34, 55, 89,
144, 233, 377, 610, 987, 1597 . . .

If you've never seen these numbers before, can you work out the pattern?

Take a minute or two before reading any further and try to work it out!

The Fibonacci sequence starts with the number 0 and the number 1. From then on, you form the next number by adding up the two previous ones. So $0 + 1 = 1$, $1 + 1 = 2$, then $1 + 2 = 3$, $2 + 3 = 5$ and so on forever.

Lots of cool patterns emerge out of the Fibonacci sequence. It's the kind of object that rewards careful attention because you can find treasures hidden within it if you're willing to dig a little. For instance, watch what happens when you take every number in the sequence and square it (that is, multiply it by itself):

0, 1, 1, 4, 9, 25, 64, 169, 441, 1156 . . .

Nothing too special? Well, look what happens if you now add up consecutive pairs of these numbers. Starting from the beginning, we get:

1, 2, 5, 13, 34, 89, 233, 610, 1597 . . .

Those look familiar! In fact, this is every second number in the Fibonacci sequence. The squares of the Fibonacci numbers do even stranger things when you add them all up from the beginning of the list rather than just taking them in pairs. That's a bit complicated to say in words, so let me illustrate with a handful of equations. See if you can spot a pattern in the following sums:

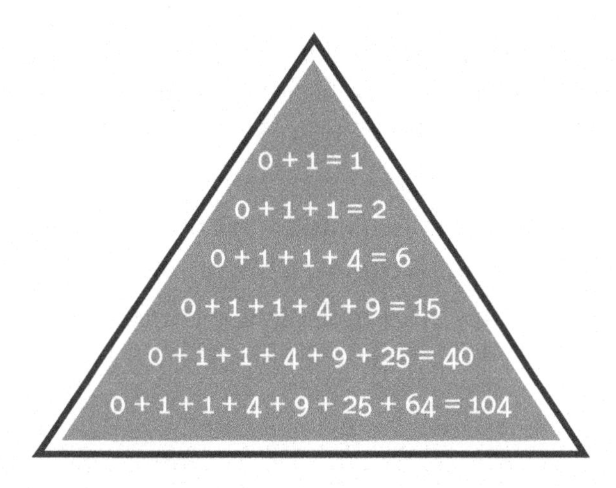

1, 2, 6, 15, 40, 104 … all right, I'll admit it. It doesn't look like there's much of a pattern here at first glance – but if I place these sums (the numbers added together) beside some products (consecutive Fibonacci numbers multiplied together), the pattern will reveal itself:

Sum	Product
0 + 1 = 1	1 x 1 = 1
0 + 1 + 1 = 2	1 x 2 = 2
0 + 1 + 1 + 4 = 6	2 x 3 = 6
0 + 1 + 1 + 4 + 9 = 15	3 x 5 = 15
0 + 1 + 1 + 4 + 9 + 25 = 40	5 x 8 = 40
0 + 1 + 1 + 4 + 9 + 25 + 64 = 104	8 x 13 = 104

That's a bit creepy, right? The real question, of course, is why this is happening. Why should the sums of the squares of the Fibonacci numbers be equal to the products of consecutive Fibonacci numbers?

To see inside this mystery and work out what's going on here, we need to see through the numbers to what they actually represent. The language we've been using throughout this discussion will help us if we pay closer attention to it. When we talk about the 'square' of a number, we often forget that we are invoking a geometric illustration: 25 is the square of 5 because it is literally the area of a square with sides of 5 units in length. So when we talk about adding up the squares of the Fibonacci numbers, what we are adding up is a series of increasingly large squares. Let me show you what it looks like.

The first sum is nothing to write home about – here's the illustration for 0 + 1:

That's followed by 0 + 1 + 1, which is still unremarkable:

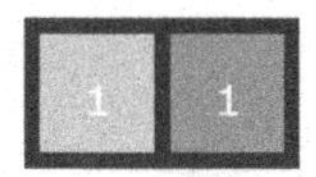

Things begin to get more interesting with the next two sums, 0 + 1 + 1 + 4 and 0 + 1 + 1 + 4 + 9.

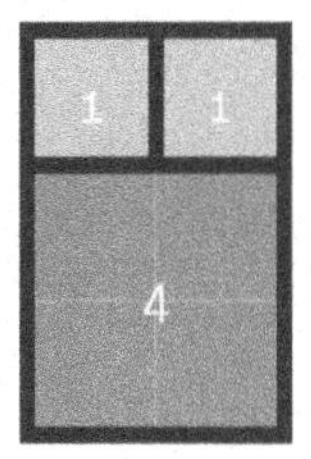

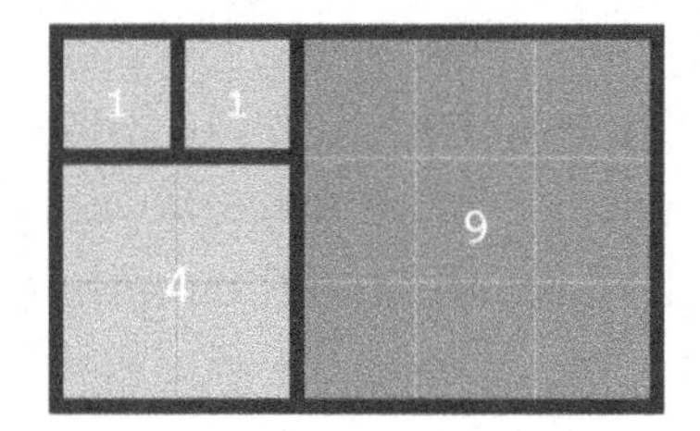

Notice how the newest square that we add at each step (marked in deep red) is always able to fit snugly onto the edge of the previous diagram. That's not a coincidence – and I wonder if you can work out why. Will it become any clearer if I illustrate the next couple of steps?

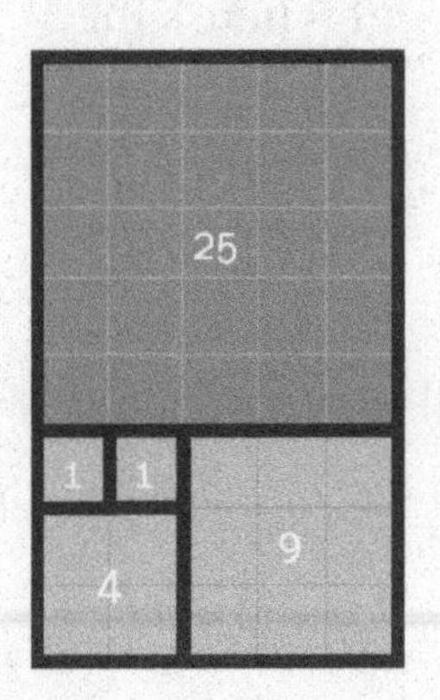

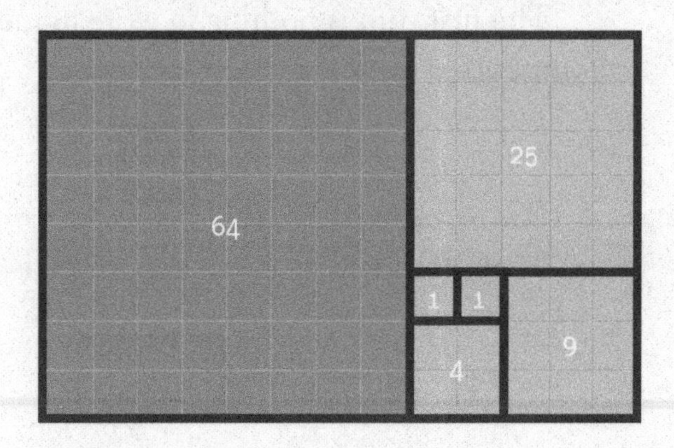

Remember those sums we were thinking about? 1, 2, 6, 15, 40 and 104? What we see here are figures that have exactly those areas. But these shapes are not just any random polygons – they are in fact perfect rectangles. This is because, as we noticed above, each new square seems to coincidentally fit onto the figure that came before it.

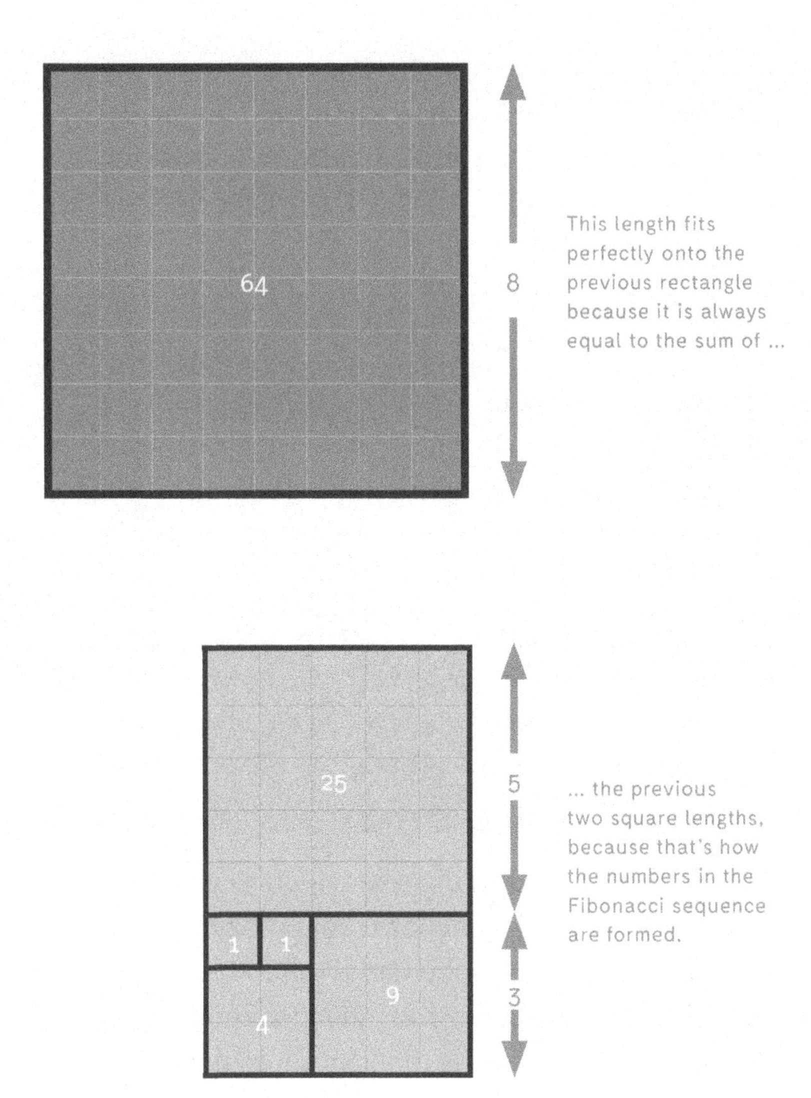

But if each new diagram is a rectangle – and not just the sum of a bunch of squares – that means we can calculate its total area in another way. The area of a rectangle is equal to its length multiplied by its breadth. But look at the length and breadth of each of these rectangles we've made: they are always consecutive Fibonacci numbers! So the last one, for instance, is 8 units tall and 13 units wide. That is why its area is not just equal to $0 + 1 + 1 + 4 + 9 + 25 + 64$, but also equal to 8×13.

By the way, if you felt an eerie sense of déjà vu as you stepped through that explanation, it might be due to how familiar those diagrams are. We saw them in the last chapter when we were talking about the Golden Ratio – each one is looking more and more like a golden rectangle. The difference is that while we previously constructed the rectangles from the outside in, this time we are building them from the inside out. This is one of the hallmarks of mathematics – the same patterns emerge again and again, no matter which direction you come from.

Okay, so that's cute. But the real reason we're here is because the Fibonacci sequence is supposed to be related in some mysterious way to the Golden Ratio, right? Well, this time instead of squaring the individual numbers in the sequence, let's do something that combines the numbers in an unusual way. This time what we'll do is divide each number by the previous term to see what happens (ignoring the zero because dividing by zero lands us in all kinds of logical troubles).

Division	Answer
1 ÷ 1	1
2 ÷ 1	2
3 ÷ 2	1.5
5 ÷ 3	1.66666666 . . .
8 ÷ 5	1.6
13 ÷ 8	1.625
21 ÷ 13	1.61538461 . . .
34 ÷ 21	1.61904761 . . .
55 ÷ 34	1.61764705 . . .
89 ÷ 55	1.61818181 . . .
144 ÷ 89	1.61797752 . . .
233 ÷ 144	1.61805555 . . .
377 ÷ 233	1.61802575 . . .

Oooooooookay – so that is definitely creepy! There's that sneaky Golden Ratio showing itself again. How does such a simple rule – adding up numbers one after the other in sequence – create something so geometrically profound? Should we rename the Fibonacci sequence the 'Golden sequence'?

Well – hold your horses. Because as you're about to see, the Fibonacci sequence is not as special as it seems. You've seen the Fibonacci numbers; now I want to introduce you to:

The Woo numbers.

The Woo numbers start off with a 19 and a 9, because my birthday is 19 September. After that, you get the next number in the sequence just the same way that the Fibonacci sequence does: you add two adjacent numbers to get the next one. Here is the beginning of the list: 19, 9, 28, 37, 65, 102, 167, 269, 436, 705, 1141, 1846, 2987, 4833 . . .

So, no big deal, right? Nothing out of the ordinary. No special properties in this list of numbers, it seems. But to be certain, let's try dividing the Woo numbers like we did with the Fibonacci numbers, just to make sure there's no funny stuff going on.

Division	Answer
9 ÷ 19	0.47368421 . . .
28 ÷ 9	3.11111111 . . .
37 ÷ 28	1.32142857 . . .
65 ÷ 37	1.75675675 . . .
102 ÷ 65	1.56923076 . . .
167 ÷ 102	1.63725490 . . .
269 ÷ 167	1.61077844 . . .
436 ÷ 269	1.62081784 . . .
705 ÷ 436	1.61697247 . . .
1141 ÷ 705	1.61843971 . . .
1846 ÷ 1141	1.61787905 . . .
2987 ÷ 1846	1.61809317 . . .
4833 ÷ 2987	1.61801138 . . .

All it takes is the calculator on your phone to test this out for your own pair of numbers. Try your birthday, or any random set of numbers. It doesn't matter how big or small they are. Every sequence produced like this eventually approaches the Golden Ratio between subsequent terms! So it seems like the Fibonacci numbers aren't so special after all. Maybe there is no list of numbers that is more 'golden' than any of the others.

That's until you meet the astonishing Lucas numbers. Édouard Lucas was a nineteenth-century French mathematician who took a keen interest in 'recreational mathematics' (where the focus is on playfulness rather than practical applications). He invented one of my favourite mathematical games of all time, 'Boxes', which you may well have played yourself by drawing dots on a page and taking turns with a friend to join the dots to see who can form the most boxes. If a player completes the fourth side of a box, they write their initial inside that box and get to draw another line.

PLAYER A = ——— **PLAYER B =** ·······

1

2

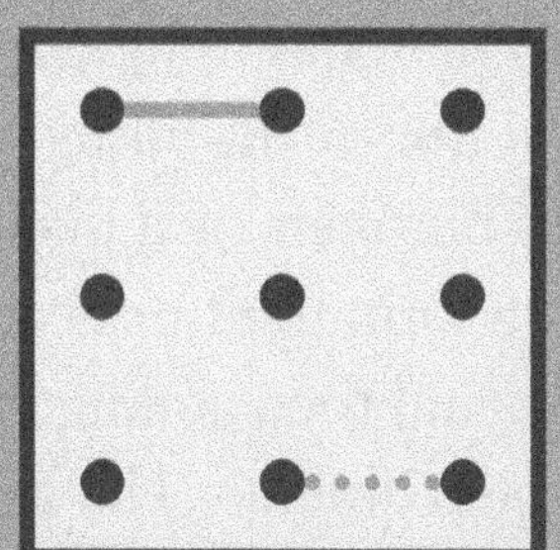

3

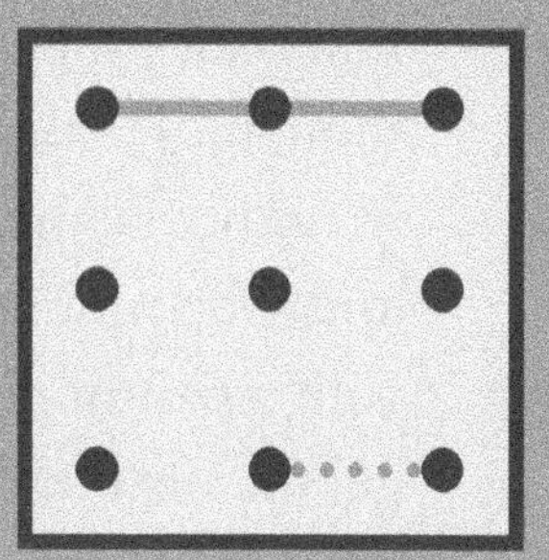

4

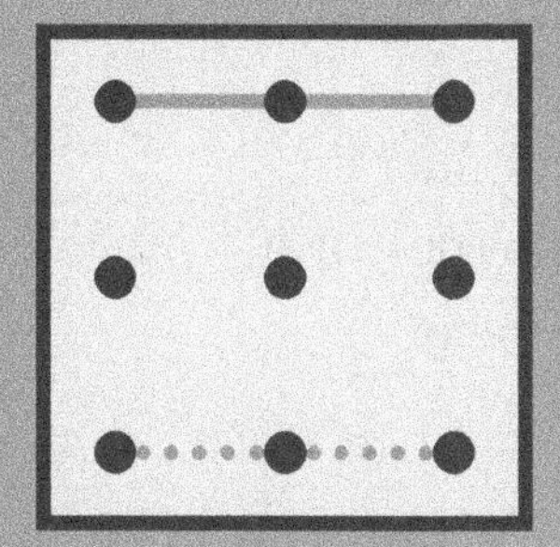

5

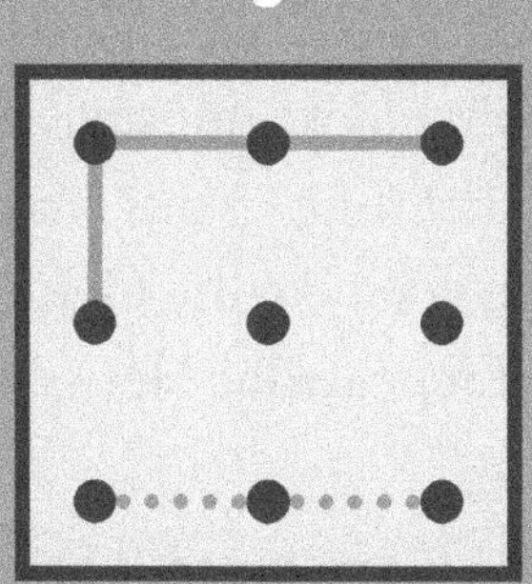

6

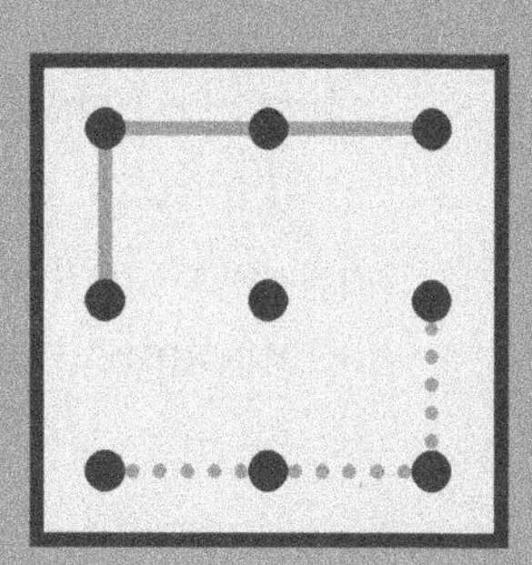

7

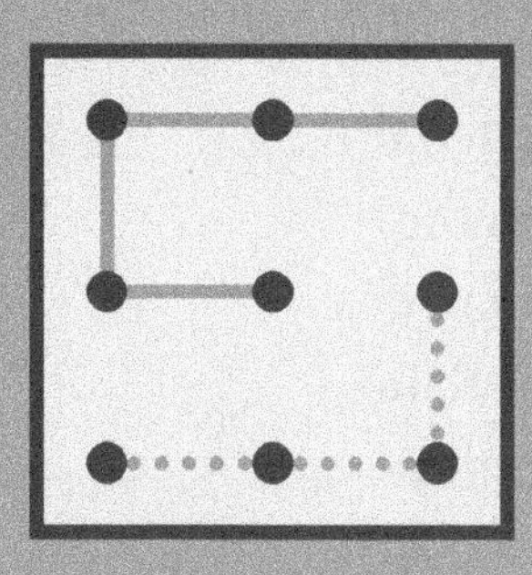

8

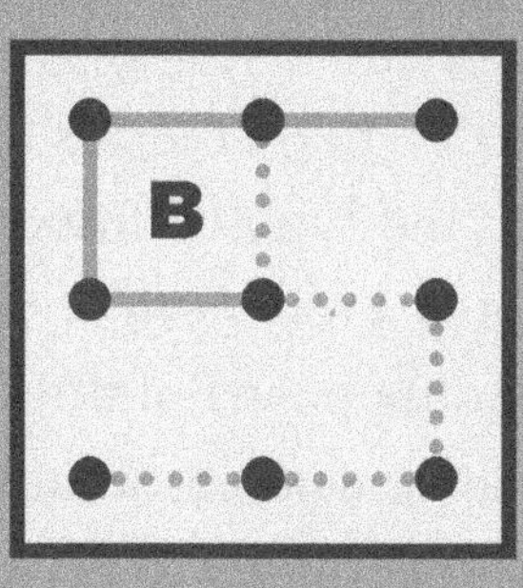

9

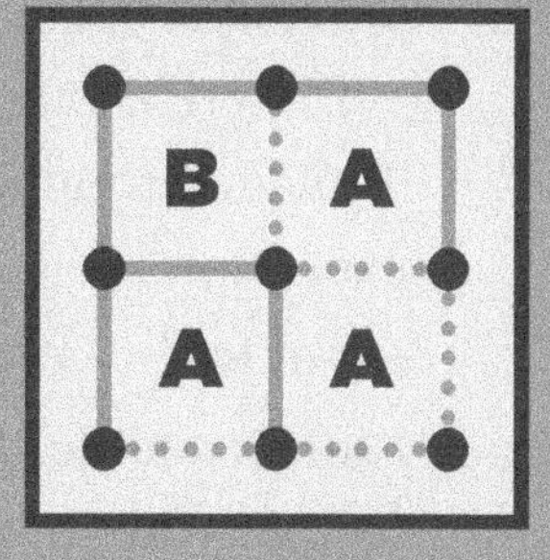

In his playful exploration, he spent a fair amount of time studying the Fibonacci numbers and came to realise that their properties were not as unique as many had once thought. He illustrated this by inventing his own sequence of numbers, the Lucas numbers. Just like the Fibonacci and Woo numbers, you form the Lucas numbers by starting with a pair of numbers and then adding them together to get the next term in the sequence. The Lucas numbers begin with 1 and 3. The sequence then proceeds like so:

1, 3, 4, 7, 11, 18, 29, 47, 76, 123, 199 . . .

Just like before, nothing looks unusual at first. But if you'll indulge me one last time in this chapter, consider the sequence created when you take the Golden Ratio (phi, or ϕ) and multiply it by itself repeatedly. Here's what I mean:

1.618, 2.618, 4.236, 6.854,
11.0902, 17.9443, 29.0344 . . .

Now have a look at these two sequences side by side in the graph opposite.

If there is one sequence that deserves to be called the 'Golden sequence', it's the Lucas numbers. With the exception of its first term, every term of the sequence is equal to the corresponding power of the Golden Ratio rounded off to the nearest whole number – and the further you go, the closer you get to the exact Lucas number. Weird!

A BRIEF NOTE: I use a piece of mathematical shorthand in the table below called a 'power' (it's also called an 'index' or an 'exponent' – which is where we get the word 'exponential'). Mathematicians always want to write things in the most efficient way possible, and they often invent all new symbols and types of notation to allow them to say things more quickly. Multiplication began as a shorthand for repeated addition: when you see 3 x 5, that's equivalent to seeing 5 + 5 + 5. That's why you literally read it out as 'three times five' (as in, 'add five, three times'). Powers are exactly the same idea, except for the next level: they began as a shorthand for repeated multiplication. That means when you see something like ϕ^3, for instance, you would read that aloud as 'ϕ to the power of 3', and what it means is $\phi \times \phi \times \phi$.

Lucas number	ϕ^n (rounded to four decimal points)
1	$\phi^1 = 1.618$
3	$\phi^2 = 2.618$
4	$\phi^3 = 4.236$
7	$\phi^4 = 6.8541$
11	$\phi^5 = 11.0902$
18	$\phi^6 = 17.9443$
29	$\phi^7 = 29.0344$
47	$\phi^8 = 46.9787$
76	$\phi^9 = 76.0132$
123	$\phi^{10} = 122.9919$
199	$\phi^{11} = 199.0050$
322	$\phi^{12} = 321.9969$
521	$\phi^{13} = 521.0019$
843	$\phi^{14} = 842.9988$
1364	$\phi^{15} = 1364.0007$
2207	$\phi^{16} = 2206.9995$
3571	$\phi^{17} = 3571.0003$
5778	$\phi^{18} = 5777.9998$
9349	$\phi^{19} = 9349.0001$

CHAPTER 09

WHAT DO YOU CALL A COMEDIAN TYING THEIR SHOELACES?

× Knot funny! ×

One of the things I love about mathematics is how good it is at solving problems.

What's the best shape for designing your satellite dish to pick up signals from deep space?

What will the temperature range be like tomorrow?

What is the fastest path from my house to the city if I need to stop at two locations along the way?

What's the right amount to charge for a cup of coffee at your cafe to make maximum profit while keeping customers happy?

How much steel reinforcement does this bridge need so that it can withstand the weight of 500 cars at peak hour?

All of these problems, and more, are solved with the aid of mathematics. More specifically, the kind of mathematics we're talking about is often called

'applied mathematics',

because it takes mathematical knowledge and techniques, then applies them to actual problems in the real world. Applied mathematics is mathematics that has been put to work. It's mathematics in overalls with its sleeves rolled up,

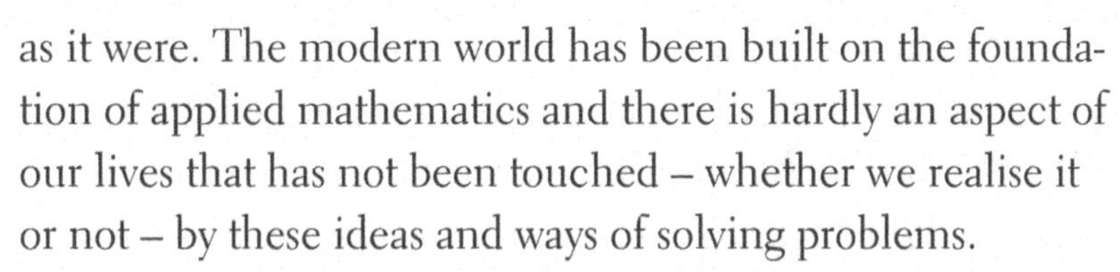

as it were. The modern world has been built on the foundation of applied mathematics and there is hardly an aspect of our lives that has not been touched – whether we realise it or not – by these ideas and ways of solving problems.

In fact, sometimes these ideas are performing at their very best when they aren't even noticed. The movie suggestion algorithm put together by Netflix, the movie and TV streaming company, is a key exemplar of this idea. If the software engineers at Netflix have done their job well, you don't notice that there's incredibly complex mathematics built into their product – you're just happy with the 'suggested films' that appear as you browse and – they hope – you spend more time watching shows.

Applied mathematics is not the only kind of mathematics, though. In this way, among others, mathematics is like music. Yes, music can be 'put to work' as an advertising jingle (to sell things) or as a national anthem (to drum up national pride). But musicians don't usually play music because it solves a problem; most people play music because they enjoy it. In mathematical terms, we call this 'pure mathematics' – it's pure in that it hasn't been touched (contaminated?) by any outside context or practical concerns. It is mathematics for its own sake, mathematics that is done for leisure.

This is mathematics in a bathrobe and fluffy slippers.

You might find it hard to imagine someone doing mathematics for fun. But, as we discuss throughout this book, that's likely because you've never been shown how broad mathematics is. A child assembling a jigsaw puzzle, a girl wrestling with a Rubik's Cube, a boy struggling with the folds to make a paper crane or a businesswoman scratching numbers into a Sudoku on her train ride home – they are all engaged in mathematical recreation of one sort or another.

Throughout history, there have been mathematicians who felt a special pride in doing mathematics without any practical purpose. They felt that applying their intellect to pure and abstract ideas was in some ways a higher pursuit than that engaged in by those who were, in a manner of speaking, using mathematics as a means to an end. The British mathematician Godfrey Hardy exemplified this kind of attitude, proudly declaring in his essay *A Mathematician's Apology*, 'No discovery of mine has made, or is likely

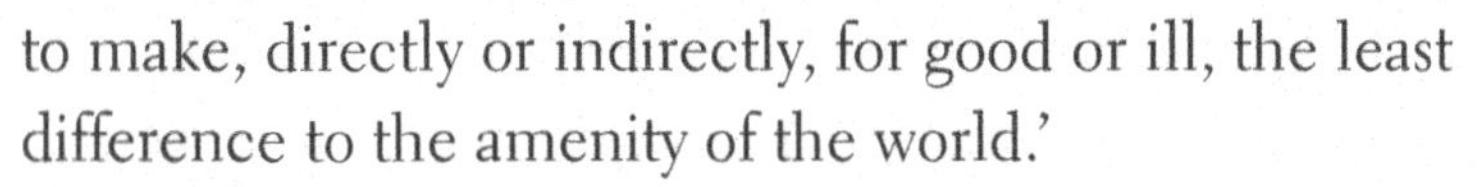

to make, directly or indirectly, for good or ill, the least difference to the amenity of the world.'

It might seem a little odd to brag about how impractical and 'useless' your life's work is, especially when the primary complaint issued by critics of mathematics education is that they don't see its relevance to everyday life. However, another of Hardy's comments in the same essay might clarify why he thought of this as a virtue. 'No one has yet discovered any warlike purpose to be served by the theory of numbers … and it seems unlikely that anyone will do so for many years.'

It's worth pointing out that while Hardy's convictions may have been very genuine, it isn't hard to argue that they were ultimately mistaken. As we will see in the next chapter, 'Unbreakable locks', some areas of mathematics that people would never dream of having practical applications – like the study of prime numbers – turn out to be incredibly useful indeed. In fact, there is a long history of mathematical curiosities conjured up by mathematicians with no view towards practicality that end up having deep and profound applications in the physical world and human society.

That said, Hardy would have liked knot theorists. Knot theory was devised in the late eighteenth century because . . . well, there wasn't really much of a 'because'. Early knot theorists were not trying to solve any problems or unlock any mysteries of the universe. They just found knots curious and sought a way to describe and classify them logically. Knots, after all, have been a part of human culture since prehistoric times: not just to tie things together, but also as a way to record information and even as a means of artistic expression. Both the Chinese and Celts have well-developed traditions of intricate knotwork that date back several centuries. Knots can even have religious or spiritual significance, such as the Borromean rings, which have been discovered and rediscovered repeatedly by cultures around the world:

It's worth noting at this point that a **mathematical 'knot'** is a little different from what you might be picturing when you think of knots. There's a good chance you're thinking right now about tying shoelaces – and it's true, objects like shoelace knots were the inspiration and starting point for some ideas in knot theory. But for reasons that will become clear a bit later on, knot theorists are interested in objects called 'closed loops' – that is, unlike a shoelace which has two free ends, we are thinking about what happens if we permanently attach those ends to each other. For that reason, the most basic kind of knot is one that, in a sense, isn't knotted at all. It's a simple ring:

The technical name for this guy is the 'unknot'. It never crosses or loops back over itself, which is, unsurprisingly, why it looks so simple. But beware – some knots may look much more complicated, but are actually just unknots in disguise. For instance, have a look at this knot:

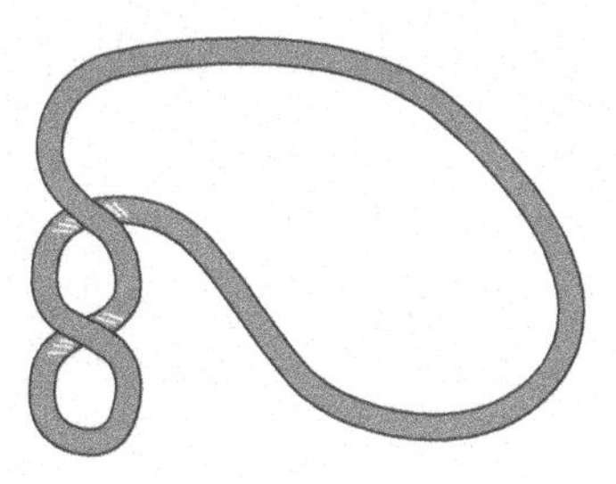

There appear to be two spots, both in the bottom-left corner, where the knot crosses over itself. However, if you take those two mini-loops and turn them over in your mind, you can easily imagine uncoiling the knot and ending up back at the unknot we saw opposite. In knot theory, that makes these two knots equivalent. As soon as you realise this, you can recognise that even monsters like this may be quite simple objects once they are tamed:

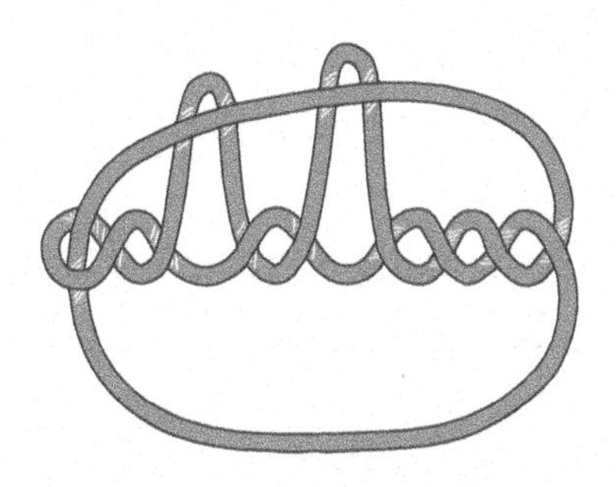

As an aside, this is why we mathematicians aren't concerned with the kinds of knots you can make with shoelaces. The whole point of shoelaces is that you can undo them and tie them again in a different way if you please. Since any 'knot' made from shoelaces can be untangled and rearranged into a different one, we say that all such knots are essentially identical when it comes to knot theory.

It doesn't take long before you come up with a knot that's genuinely different, though. Consider the following design:

This knot is called the 'trefoil'. It's named after the three-leaf clover plants in the Trifolium genus, whose leaves resemble this knot as you see it above. You can get a pipe cleaner or a piece of string yourself and try to replicate it, then fuse the ends together. No matter how hard you try, you'll never be able to untwist it and turn it into an unknot. It's another beast entirely.

I encourage you to actually do this. Try manipulating the knot so it looks different from the diagram above, and then sketch out your knot on a piece of paper. If you do it a few times, you may notice that no matter how you wiggle or manipulate the various parts of the knot, it will always overlap or cross over itself at least three times:

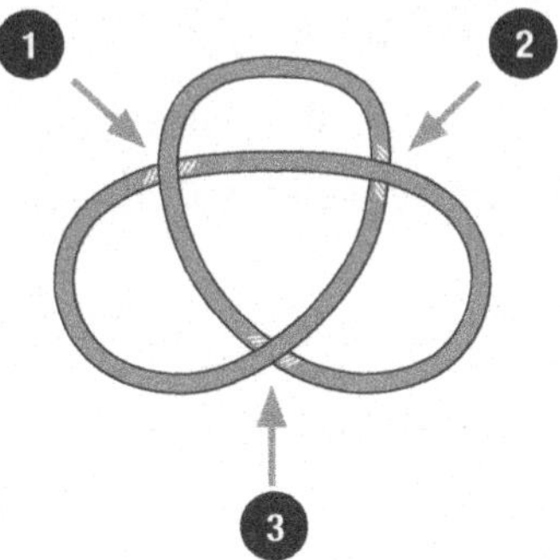

Here is the 'classic' presentation of the trefoil knot, with the three crossings highlighted. And here is what happened when I had a go at twisting and turning it a few different ways:

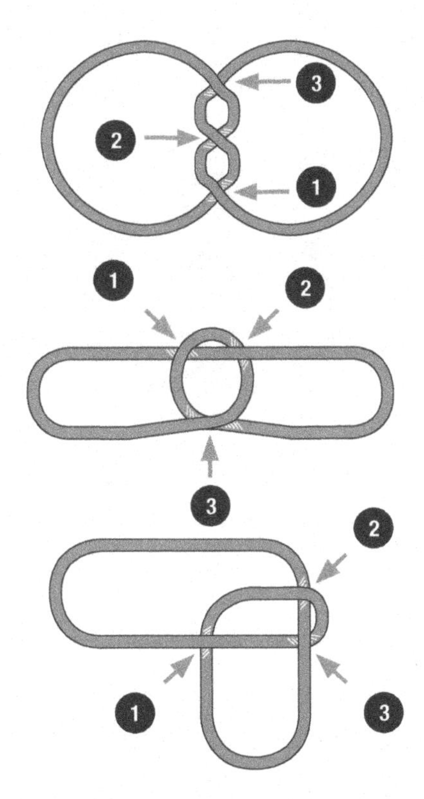

The three crossings are always clearly visible. I can obviously add more crossings by taking a section of the loop and twisting it over itself, but no matter how I deform the knot I will always have at least three. That is in fact the essential characteristic of any knot: the number of crossings that it has.

* The unknot has zero crossings.
* Curiously, it is impossible to make a knot with just one or two crossings (such knots can always be untangled to become the unknot again).
* The trefoil knot has three crossings.

The knot with four crossings is called the 'figure eight knot', and you can see why when you have a look at its classic representation:

As we add more crossings, knots become harder and harder to distinguish from each other. For instance, the two following knots look quite different:

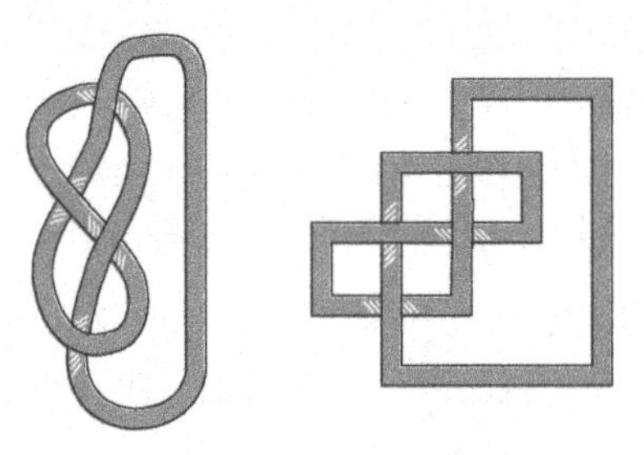

But a careful count reveals that they are really the same as the figure eight knot, because they each have four crossings and can be deformed neatly into the same configuration.

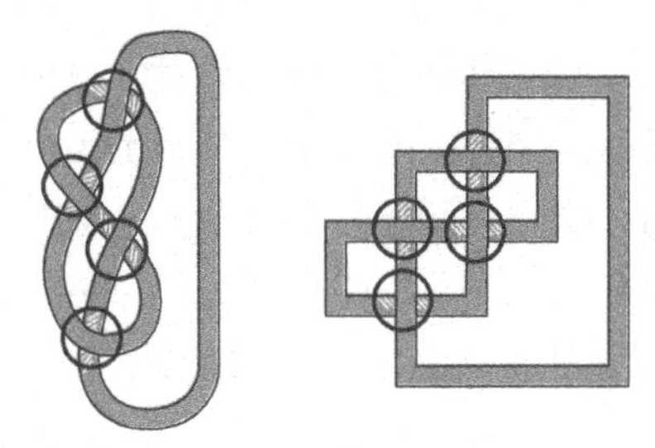

But that last phrase is very important. If you can't deform one knot into another without resorting to cutting and reattaching the loop, then the two knots are genuinely different, even if they have the same number of crossings. See the cinquefoil (left) and three-twist knob (right) below:

These knots each have exactly five crossings, but it is impossible to twist one into the other. So while there may only be a single kind of knot with zero crossings, three crossings or four crossings, there are two different kinds of knot with five crossings. There are three knots with six crossings, seven knots with seven crossings, 21 knots with eight crossings, 49 knots with nine crossings, and a whopping 165 knots with ten crossings.

We began this chapter talking about the difference between applied mathematics – the kind that is invented to solve problems that we encounter in the real world – and pure mathematics, which is studied completely for its own sake. Knot theory seems like it belongs pretty squarely in the latter camp – who, apart from mathematicians and Scouts earning badges for the number and difficulty of knots they can tie, would really be interested in these things anyway?

The ultimate twist of fate is that knot theory is personally important to every single living thing on the planet. Each cell in your body is full of knots that make you who you are. I'm referring to deoxyribonucleic acid, known more commonly by its acronym, DNA.

DNA carries the genetic instructions that govern the growth and functioning of all known organisms (and even some non-living ones, depending on how you define viruses). DNA is essentially a code made up of organic molecules that are strung together in a very specific order. Just as the order of letters in a word defines that word and distinguishes it from others, so the order of molecules in DNA is its defining quality.

Human beings are complicated creatures, and so our genetic code is extremely long so as to be able to carry all the information required. While the English alphabet may have 26 characters, the 'alphabet' of DNA contains only four: nitrogen-based molecules called cytosine, guanine, adenine and thymine. It takes three billion pairs of these molecules to make you, and all three billion pairs are found in each of your cells.

Those special molecules, called 'bases', are small – but three billion of anything is going to add up pretty fast. If you took the DNA from just a single cell in your body and stretched it all the way out, it would be about two metres long. But according to our best research, there are about 37 *trillion* of those cells in an average human body. So if you lined up the total amount of DNA in your body, it would be over 74,000,000,000 kilometres long from end to end. How far is that, you ask? To give you a sense of scale, it's about as far as you would travel if you took a round trip from here to the sun – 250 times.

That's how much DNA is in your body. And yet it's actually stored in a space smaller than the human eye can see. How is this possible? Answer: knots. Your DNA is coiled up in knots that are just like the ones we've been looking at throughout this chapter.

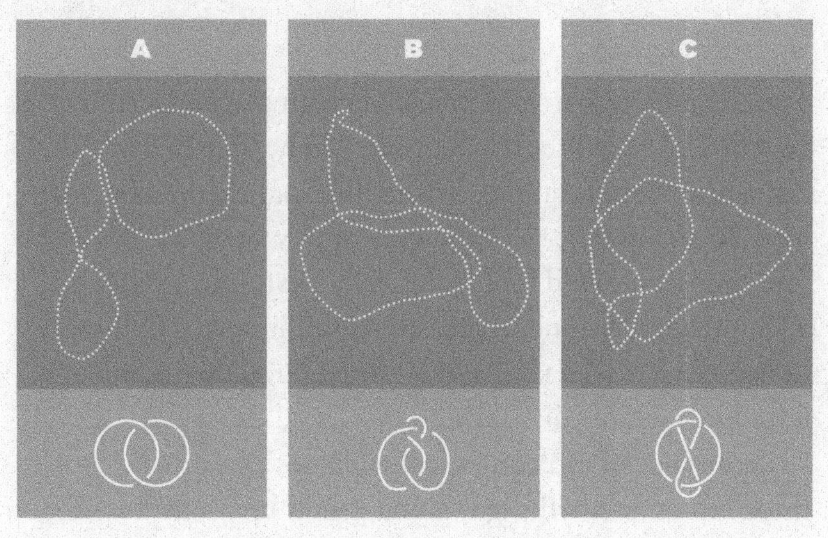

A representation of your DNA, explained in knot-terms

In fact, some of the enzymes in your body exist for the sole purpose of untying and re-attaching DNA molecules to change them from one kind of knot into another:

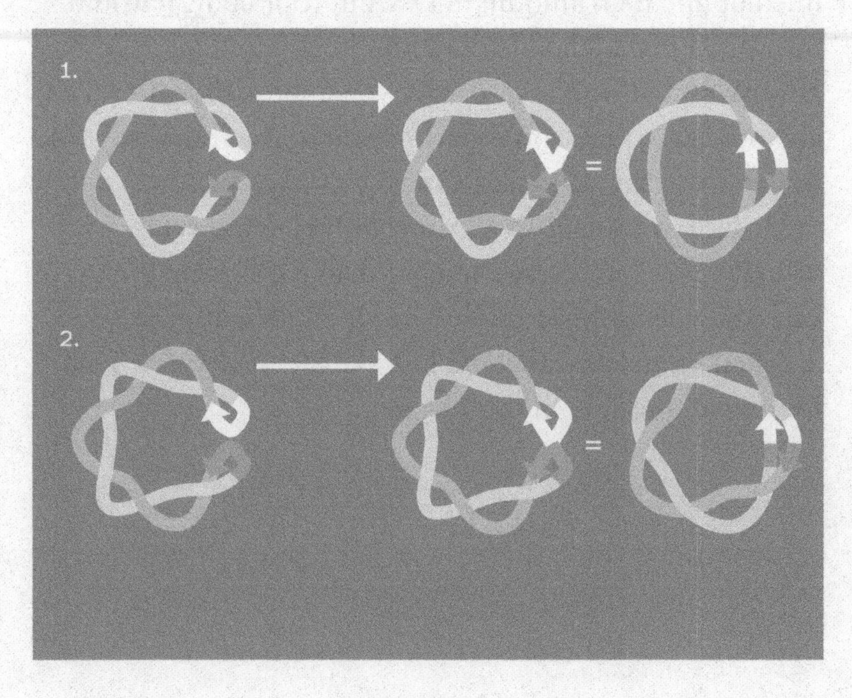

Do you realise what this means?

Right now, your cells are tying and untying knots to

keep you alive.

Your continued existence depends on

knot theory.

For me, it doesn't get any deeper than this. Why does mathematics matter? It matters because it holds the secrets to helping us understand the mysteries of the universe – even mysteries as profound as the genetic code that animates every living thing on the planet.

CHAPTER 10

UNBREAKABLE LOCKS

Telephone (also known as 'Chinese whispers') is a timeless game because it is both simple and endlessly entertaining. It's always funny to see how a seemingly straightforward phrase can morph over the course of a very small number of repetitions into something completely unrecognisable.

It's a little-known fact that the modern internet bears a remarkable similarity to this amusing party game. When you do something as simple as opening an app on your phone, your phone whispers by radio to a nearby tower, which whispers through underground cables to a local signal exchange, which whispers to a server belonging to an internet service provider. The chain of whispers continues as your phone's message travels out of the country and along the sea floor to, most likely, the United States. At last, the signal arrives at a receiving computer which reads it: 'Send me the home page of Google!' The receiver dutifully follows the instruction, whispering back a set of data that returns to your phone, most likely on a completely different path from before. This round trip, over 24,000 kilometres long, takes less than a fifth of a second.

When comparing the internet to a game of Telephone, it's natural to think – especially over such a distance – that the challenge is in ensuring that the message received is identical to the message sent. This is indeed a problem, and mathematics happily steps into the breach to solve it. There are thousands of mathematical ways to ensure that messages arrive intact. One of the simplest to understand is called the 'check digit' – here's an example of how it can work.

Suppose you wanted to send the eight-digit number 26101949 across to the other side of the world. When you hit send, it's a little bit like dropping your message into a post office. The office will inspect the letter, put a stamp on it, and send it on its way. The difference here is that rather than a stamp, your computer will attach a ninth digit – the 'check digit' – onto the end of your number and then send that. Here's how your computer works out what extra digit to include:

1. It adds up all the digits in the message: 2 + 6 + 1 + 0 + 1 + 9 + 4 + 9 = 32.
2. It takes this total and then subtracts 10 until the number is less than 10: 32 - 10 = 22, 22 - 10 = 12, 12 - 10 = 2.
3. The final number (in this case, 2) is the check digit.

So instead of sending 26101949, your computer sends 261019492. The receiving computer, on the other side of the world, knows what you did. You would have agreed upon a set of rules (or 'protocol') for how to do things before you even started communicating. If the message arrived exactly as we sent it, then the receiver will go through the same process as you did originally (adding up all the digits) and will arrive at a value for the check digit, which it believes should be 2. Since your check digit agrees with theirs, it concludes that the message arrived accurately.

Suppose something went wrong. If one of the computers in the chain of whispers corrupted the message by accident and sent 231019492 instead, then the receiving computer would go through the following steps:

1. It adds up all the digits in the message: 2 + 3 + 1 + 0 + 1 + 9 + 4 + 9 = 29.
2. It takes this total and then subtracts 10 until the number is less than 10: 29 - 10 = 19, 19 - 10 = 9.
3. This means the receiving computer expects that your check digit should be 9, but the check digit it received is 2. This means there must have been an error somewhere along the way.

This particular set of steps (called an 'algorithm') for making and verifying the check digit is very simple and very quick for computers to calculate, but the hidden cost is that there are many errors that will slip through the cracks. For instance, if any digits are moved into a different place (e.g., 61021994) – which is called a transposition error – this method will produce the same check digit even though the message is drastically different. The computer will also fail to notice when multiple digits change if the changes cancel each other out: for instance, 35101949 will still produce a check digit of 2 because 3 + 5 = 8 and 2 + 6 = 8 as well. More sophisticated algorithms will capture more errors but are harder to understand and more work for computers to calculate.

This checking process actually happens not just at the end of the journey, but at every step of the way so that if an error is detected it can be fixed up immediately rather than going all the way back to the start of the chain. It's almost like each person whispers back when it is their turn, 'Is this what you really said to me?'

So that tells us a bit about how our answer arrives intact. But this then raises a much bigger question: if the internet consists of a self-checking and self-fixing chain of whispers, that means every server that receives our message and passes it on has – if not permanently, at least for a window of time – an accurate copy of the exact message they received before passing it onto the next server. If the exact message you sent was your credit card number, then does that mean hundreds and thousands of computers connected to the internet now have a perfect copy of your precious financial details? How is it that your bank account hasn't been emptied out within five minutes of you performing your first transaction on the internet?

The answer – which not only keeps you personally safe, but also forms the foundation for a multi-trillion-dollar global economy – rests in some ingenious mathematics. Before we get to the nuts and bolts of how mathematics keeps your credit card number safe, though, we need to understand something fundamental about how to send messages and keep them safe.

People have been transmitting secret messages for centuries. While one could hope that a fast and trustworthy messenger could securely deliver a message to your intended recipient without being intercepted, it is incredibly risky to entrust the security of your secrets to a single person or means of transport. In times of war, when secret communication is of the utmost importance, messengers are prime targets for capture. And, once the era of radio communication began, anyone with a decent radio could pick up the signals being broadcast nearby, whether they were the intended audience or not.

This is what led to the rise of what we now call encryption – the use of a code to scramble up messages so that they appear to be gibberish to anyone without the proper knowledge to decipher them. You could compare this to sending your message out inside a locked box, to which only you and your intended recipient have the keys. In this case, the security of the message depends entirely on keeping the keys secret; if someone could make a copy of your key, your messages would be just as easily read by them as by the intended recipient. Since the keys have to be kept secure, this strategy is called 'private key encryption'.

Here's an example of what it might look like in mathematical terms. Think back to our message from before, 26101949. A simple numerical key might be the number 5: we could use it by simply adding 5 to each digit individually and replacing each digit with its new 'encrypted' version. For the purposes of this particular encryption scheme, we will simply ignore tens if they appear when we add 5; so when we encrypt 9, we will calculate $9 + 5 = 14$ and replace the 9 with a 4. This means that the encrypted message will be 71656494. Since this technique 'transposes' one number directly onto another just like a musician transposes a song from one musical key into another, it's called a 'transposition cipher'.

While this is an improvement on sending the original message without any kind of encryption, transposition ciphers have some significant limitations. For starters, if your numbers represent letters in the alphabet and you are sending a message that consists of ordinary words in a common language like English, your encrypted message is almost as easily read as the 'plain text' original – especially if your message is quite long. Just like infantry gave way to cavalry and naval cannons were superseded by cruise missiles, there has been (and still is) a mathematical arms race between the cryptographers trying to keep messages secret and the cryptanalysts who are trying to crack them.

The simple transposition cipher was defeated by an equally simple statistical tool called frequency analysis. You see, while numbers in a message may be random, letters in words are not random at all. Some letters (like the vowels) are very common, while others hardly ever come up in normal language (like 'q' and 'z'). You might have heard that 'e' is the most common letter in the English language. If letters were randomly distributed, you would expect each one to come up about 3.8% of the time – but, according to Professor Robert Lewand in his book *Cryptological Mathematics*, 'e' turns up more than three times as often, making up a whopping 13% of most normal sentences in English. The next most frequent letter is 't', which turns up around 9% of the time. All of these numbers have been reasonably well established, and the typical distribution of letters in every language is known:

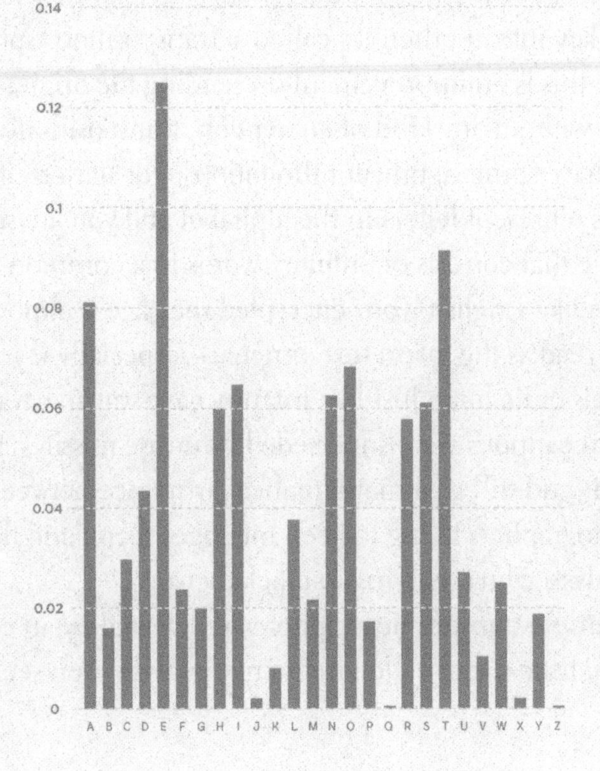

The longer a message is, the closer it conforms to this established distribution. The reason why a shorter message is more likely to diverge from this pattern is, again, a matter of the numbers. A short message like the phrase 'I am a Zulu warrior' has a distribution that looks like this:

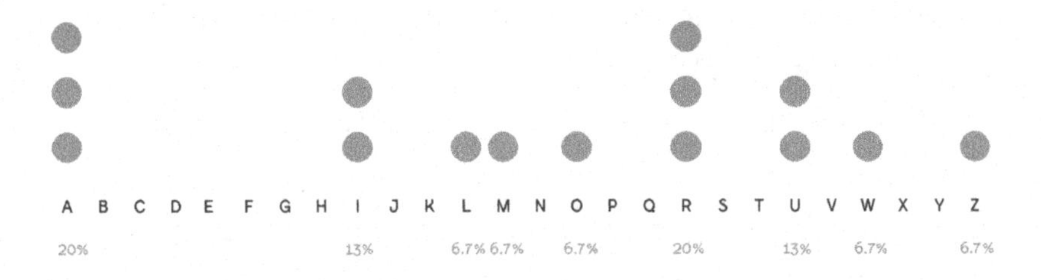

Its brevity allows uncommon words and letters to have an uncharacteristically large impact on the overall distribution. It doesn't even include a single use of the most common letter in the language! But if a message is long, this becomes less and less likely. In that case, you can scrutinise the encrypted message and check to see if the distribution of its characters matches a known distribution. If a letter or number is always transposed to the same character, then an enemy can simply match the most common letters in the encrypted message with the most common letters in the expected language to crack the code. The 'secret' message becomes an open book in the hands of a cryptanalyst.

There are ways to overcome this shortcoming. One of the most well-known was the Enigma device employed by the Nazis during the Second World War. Through a series of circuits and cogs that could be arranged in millions of different combinations, the Enigma cleverly avoided the frequency trap by replacing one character with a different letter or number each time it was encrypted. It was an inventive approach that was originally designed to keep financial transactions confidential, but the German military recognised the device's potential. They co-opted its design and upgraded it to suit their own purposes, and were thereby able to send messages that were kept secure from Allied listening outposts.

While the Enigma represented a vast improvement over its predecessors, it had a fatal flaw: it was built on the same foundation as the simple transposition cipher – a private key. All German radio officers were supplied with a code book that prescribed a series of numerical settings for their Enigma device. Each day they would change their codes, frustrating the efforts of their British counterparts who were trying to decipher their messages. The British mathematician Alan Turing led a team that eventually cracked the Enigma with a device of their own – known as the Bombe – which is a fascinating story in itself. But their success was only possible because all they needed to do was discover each day's private key, which rendered every message sent that day transparent.

A private key, even a sophisticated one, won't work in the age of the internet. If we return to the example of the chain of whispers we were thinking about earlier, we can see that there's no way for the message sender to get the private key to the other end of the chain without physically meeting that recipient. The Germans overcame this by creating their code book so that radio officers would have their keys with them wherever they went. But this, too, relied on the fact that there was a physical interaction between sender and recipient at some point in the past. Nothing like this exists between you and the internet server you're sending your credit card number to. That's why the internet isn't built on private keys: it's built on public keys.

'Public keys' sounds like a contradiction in terms. How secure is a house if its keys are kept in public? However, public key encryption works on a whole different premise from private key encryption. Remember, this is how private key encryption works:

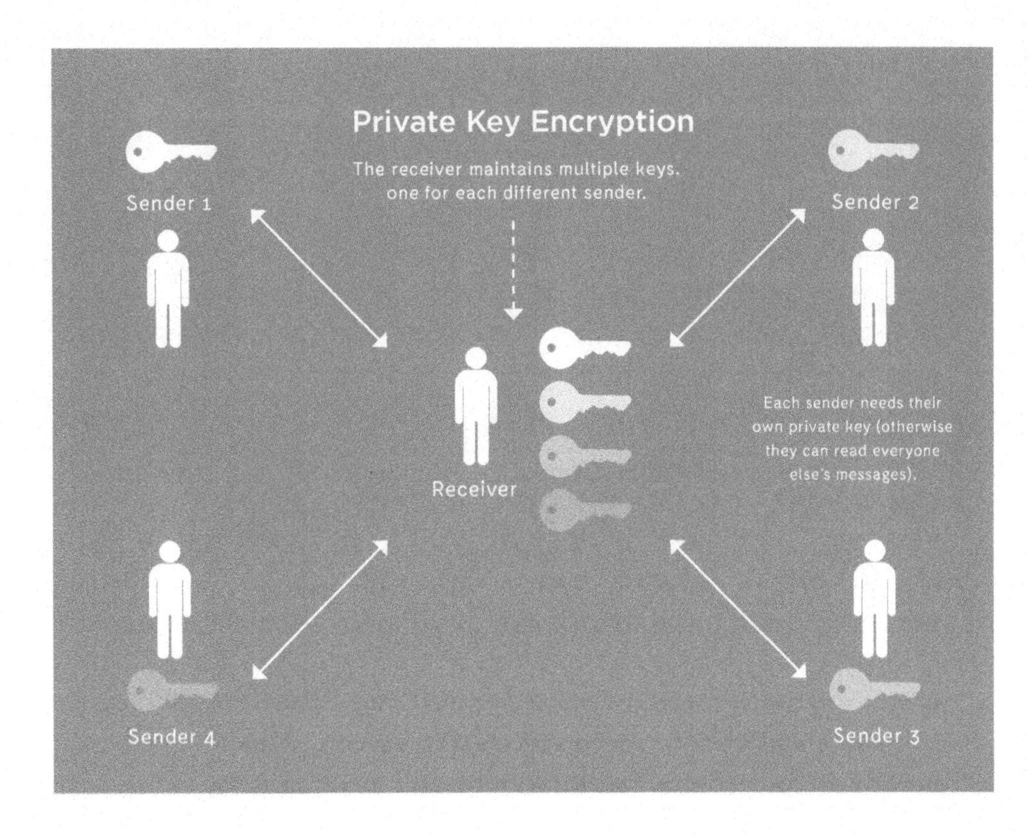

Notice that both sender and receiver need to have the same key, and the same key is used to lock and unlock their messages. This is why private key encryption is sometimes called 'symmetric encryption', since both sides are equal in the transaction.

Public key encryption, on the other hand, is totally different. The 'keys' it gives out aren't like keys at all – they are more like padlocks.

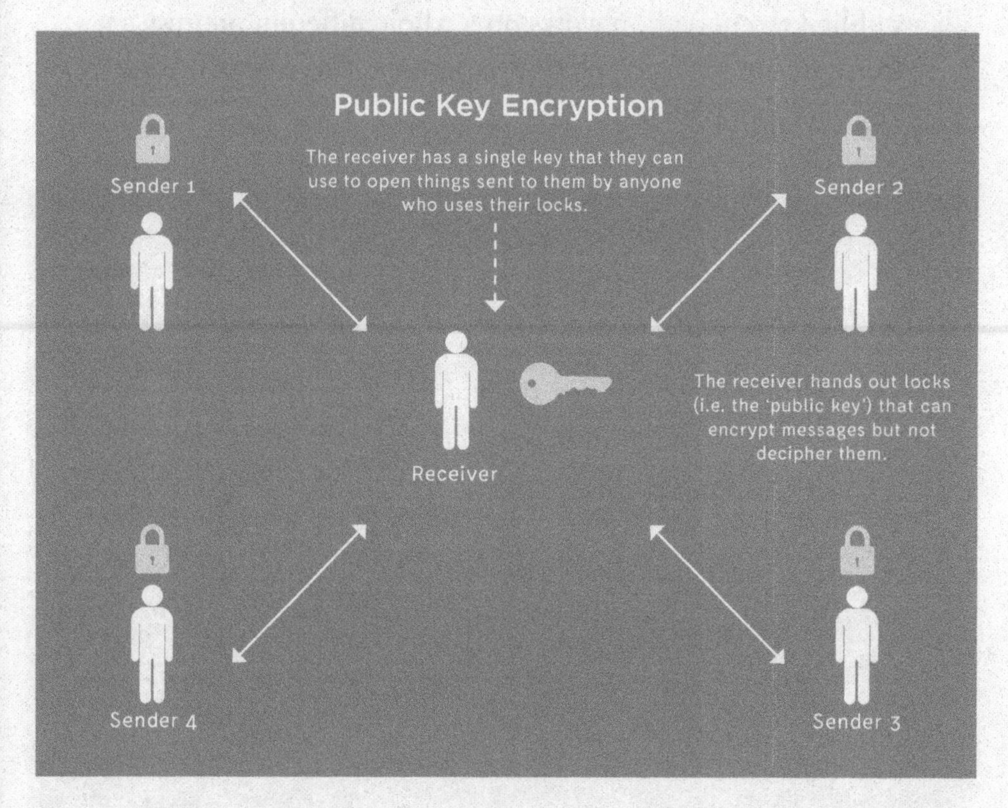

Padlocks are special. You can lock something up with a padlock even if you don't have the key. That means you can lock something but not be able to open it again.

In cryptography terms, that means you can encrypt a message but not be able to decipher it again with the same information. The receiver makes their padlocks public so anyone can lock up their message and transmit it – but only the receiver has the key to open the padlocks. Since sender and receiver no longer need the same kind of information here, public key encryption is also known as 'asymmetric encryption' – it's not balanced like private key encryption is.

This imbalance is the secret sauce of what makes public key encryption work. It requires what mathematicians call a 'trapdoor function'. As the name suggests, trapdoors are easy to fall down into but difficult to climb back out of. Trapdoor functions are easy to calculate in one direction but difficult in the other. What might this look like? We're now going to meet one of the most important kinds of numbers in existence – the prime numbers.

Children learn about prime numbers at a fairly young age, so you might still have a definition for these tucked away in your memory. Lots of people remember them as the whole numbers that can only be divided evenly by themselves and the number 1. For instance, 7 is prime because you can't divide it evenly by anything but 7 or 1 ($7 \div 7 = 1$ and $7 \div 1 = 7$). The number 6, on the other hand, is not prime because you can also divide it neatly by 2 or 3 ($6 \div 2 = 3$ and $6 \div 3 = 2$).

One of the upshots of this is that if you have a prime number of objects, it's impossible to share them fairly in a group unless you have exactly the same number of people as objects. This, by the way, is my theory of why there are exactly 11 Tim Tams in a pack: 11 is a prime number, so unless there are exactly 11 people living in your house (or people are happy to break Tim Tams in half to share – but honestly, who would do that?), it's mathematically impossible for everyone to eat the same number of Tim Tams. This of course results in an argument as you near the end of the pack and ensures that you buy another pack ('Well, you had one more than me last time!'). Pure marketing genius.

★ Prime numbers are the elements of the mathematical universe. ★ ★

Every substance in the cosmos can be made by combining a unique set of elements, and every number in existence can be made by combining a unique set of prime numbers. (This wonderful little insight is so mathematically important that it gets a fancy name, the Fundamental Theorem of Arithmetic – if you're curious about this, take a look at page 223.) And just like elements, it's easy to mix up the primes but hard to separate them from each other once they're tangled.

For instance, it's possible to multiply the prime numbers 31 and 59 with relative ease. But what are the prime numbers that are multiplied together to give 1349? Not so easy! (The answer is 19 and 71, by the way.) This kind of problem isn't just hard for our squishy brains; even insanely powerful computers find this problem (called prime factorisation) incredibly time-consuming as the numbers get larger and larger. Multiplying is easy, but factorising is hard: it's a trapdoor.

Here's how all of this links back to encryption. Websites that want to receive messages securely give out public keys to anyone who asks. The public key is an enormous number – hundreds of digits long – that anyone can use to encrypt their message. Once you've done that, the padlock is on. No one can open it back up again, not even you, without the secret key. But the secret key is the pair of prime numbers that were multiplied together to make the public key (like 19 and 71, only thousands of times larger). Factorising the public key to get these prime numbers is an intractably long task: some of these keys are so long that it would take current computers more time than the age of the observable universe to work out the answer. So the receiving website, which made the original public key using those primes, has a secure way to receive any messages it likes. No chain of whispers can endanger it.

CHAPTER 11

LOCATION, LOCATION, LOCATION

For an unremarkable patch of lawn in Tuscany, this particular location is getting a lot of attention despite how hot it is at this time of year. There are dozens of tourists scattered around, but not really paying attention to one another. It isn't the number or density of foreigners that is unusual about this scene, though; it's what they're doing.

One woman reaches an arm out as if she's propping something up. A tour group lines up and leans in unison. A young girl points a finger like she's pressing a button. All of them, as well as numerous other tourists striking elaborate poses, are having a photo taken. Where are they? They're in front of the Leaning Tower of Pisa, gathering obligatory photographic evidence of their tourist trip.

It's all in good fun. Some of the photos that people take in front of the tower are truly inventive. The really good ones look quite convincing, even when a part of our minds knows we're being tricked. But these pictures raise a genuine question: how does our mind work out how far away an object is? How does our brain work out the difference between an object being big and being close?

A huge part of the human brain is devoted to visual processing. Using our eyes to calculate distance was an important and useful skill to our ancestors, who had to rapidly work out whether fight or flight was the best option when faced with a threat, such as a tiger in the jungle. So the brain has developed a huge number of tricks to take advantage of our binocular vision, and evidently nature has figured this is a very useful quality throughout the animal kingdom: the world abounds with two-eyed creatures. Some of those tricks are very sophisticated, drawing on subtle lighting cues, the interpretation of motion or recognition of specific shapes that occur frequently in our environments. But one of the most straightforward and elegant methods that the brain uses is based on some very simple geometry.

Take a second to look up at your surroundings. Inspect the objects near you as well as the ones further away. The first thing I want you to notice is that you are most likely to be experiencing a phenomenon called singleness of vision – namely, when you look around, you perceive one image, one view, and not two. You might not find that unusual, but your brain is actually doing backflips to make this happen – because, after all, your eyes are dutifully sending two different images to your brain, and your brain has the wonderful task of knitting them into one seamless whole.

If you don't believe me, hold your right index finger up

a short distance from your eyes. Focus on it closely and observe its features. Now close one eye and look again – then swap to the other eye. If you swap back and forth you'll notice that each eye is giving your brain a different image. In fact, position your finger just right and you'll notice that your right eye can see your fingernail while your left eye only sees your fingerprint. Your eyes have no choice but to give different images because they have different perspectives – they are necessarily separated on your face.

Now you're ready to experience the beautiful mathematical trick behind how your two eyes help you perceive distance. Hold your finger up in front of your face again, but this time focus on something further away rather than on the finger itself. If you're indoors, perhaps concentrate on the opposite wall. Now repeat the trick we tried before: close one eye, then swap to the other. As you go back and forth, you should notice something else happening: your finger will appear to jump back and forth from left to right, as if it were teleporting from one spot to the other every time you blink.

Yet you know that no movement is happening. In fact, if you keep your sight focused on something far away but remain aware of where your finger is positioned in your vision, you should be able to see your finger in two places at once!

Why don't you notice this all the time? There are at least two reasons. Firstly, you aren't usually looking for visual oddities, so they escape your attention. But secondly, your brain is taking those two images – one from each eye – and stitching them into a single three-dimensional scene, every moment of every day. And it is taking advantage of the fact that the images are different to help your brain identify which objects are close and which are far away.

Here's how it works. If you play around with this eye-finger trick for a little while, you'll realise that the closer your finger is to your face, the greater the discrepancy between your finger's position in the left-eye image and in the right-eye image. Your brain knows the discrepancy not just for your finger, but for every single object in the scene in front of you. The smaller the discrepancy, the further away the object must be, since your left and right eyes are producing more or less the same image for that object. But as the discrepancy increases, your brain recognises that objects are positioned closer and closer to you.

This is why trick photos at the Leaning Tower are so impressive, even when we know they look ridiculous in person. Placing your friend and the Tower on a flat image removes the binocular cue that helps us to tell that one object is further away than the other. Our left and right eyes will provide the same image of the photo to the brain because they have little choice. We know it's wrong, but our brain is confused just enough to be fooled – at least a little.

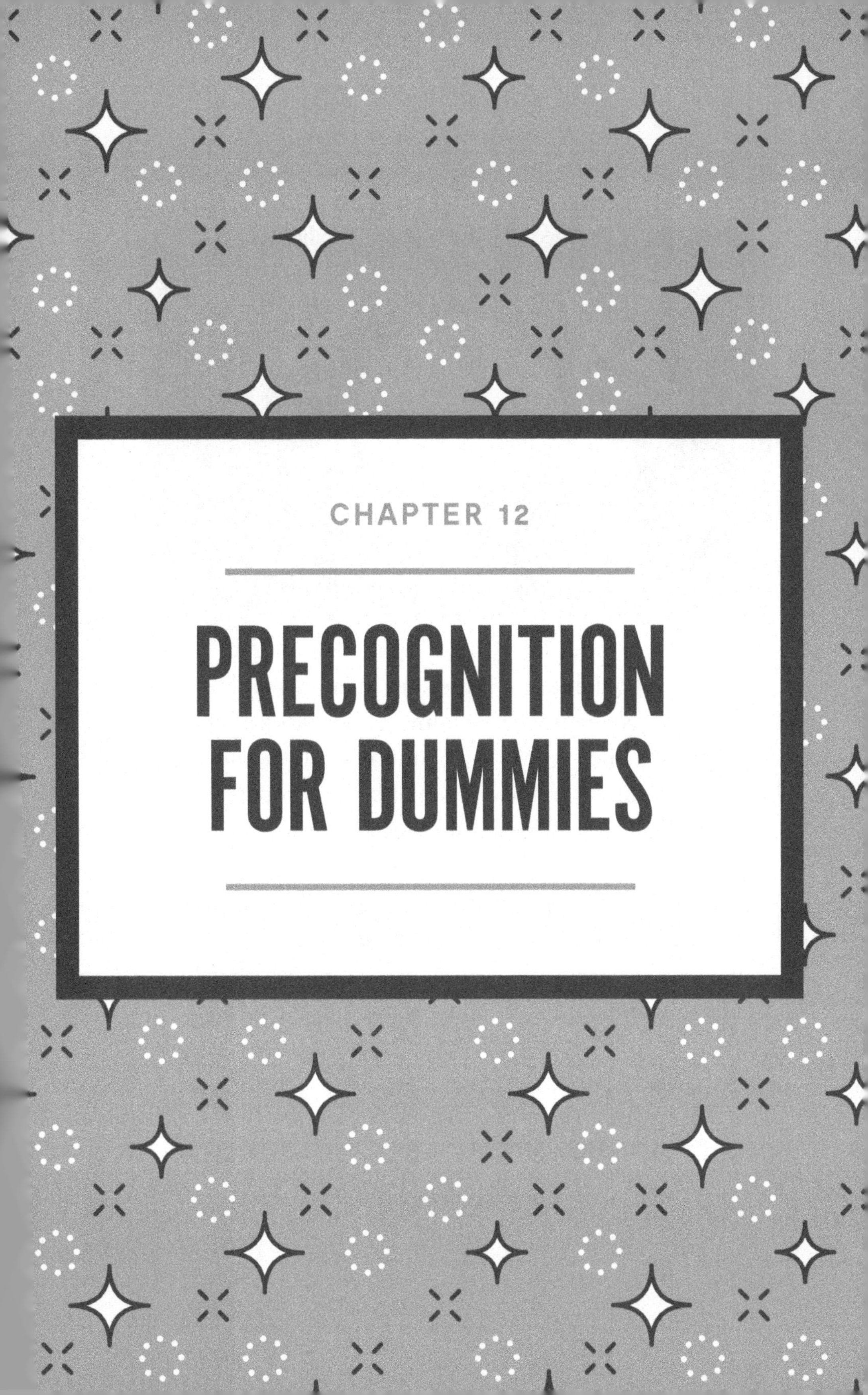

CHAPTER 12

PRECOGNITION FOR DUMMIES

Seeing into the future is the stuff of fantasy and science fiction. For some reason, almost every story involving precognition ends up being a cautionary tale in one way or another – whether in the conceit of law enforcement when they create a controversial 'Precrime Department' (as in *Minority Report*) or in the tragic twists and turns that result from a self-fulfilling prophecy (*Oedipus Rex* for those who are classically inclined, or *Kung Fu Panda* for the younger demographic). But very few of these stories reckon with the fact that humanity can already see into the future.

You don't need a crystal ball or a prophetic scroll; all you need is mathematics.

While the world sometimes appears to be governed by random or unpredictable events, the mathematical fields of probability and statistics have shown that many realities can be predicted with surprising accuracy. This is what Sir Francis Galton, the nineteenth-century English statistician, was seeking to demonstrate when he designed what was then called a Galton board but came to be known as a quincunx.

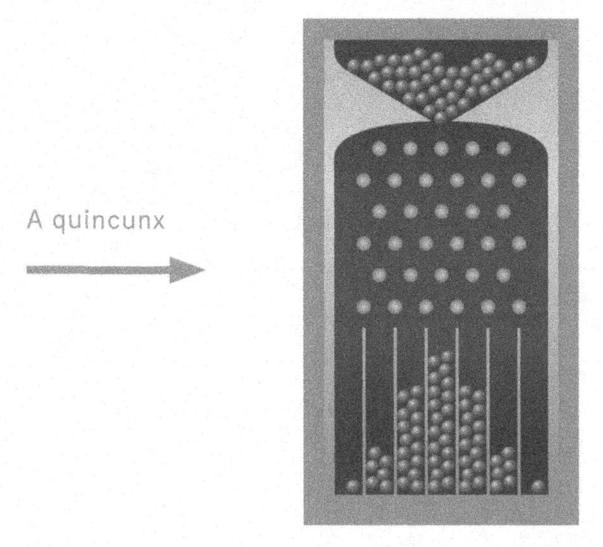

A quincunx

The quincunx is a simple invention, essentially an upright triangle with many pins nailed into its face pointed directly away from the board. A container stores balls at the top of the board, ready to be released to fall under the force of gravity. The pins are equidistant from each other and positioned such that when balls collide with them, they have an equal chance of falling to the left or the right. Eventually the balls make it to the bottom of the board and come to rest in one of the waiting receptacles.

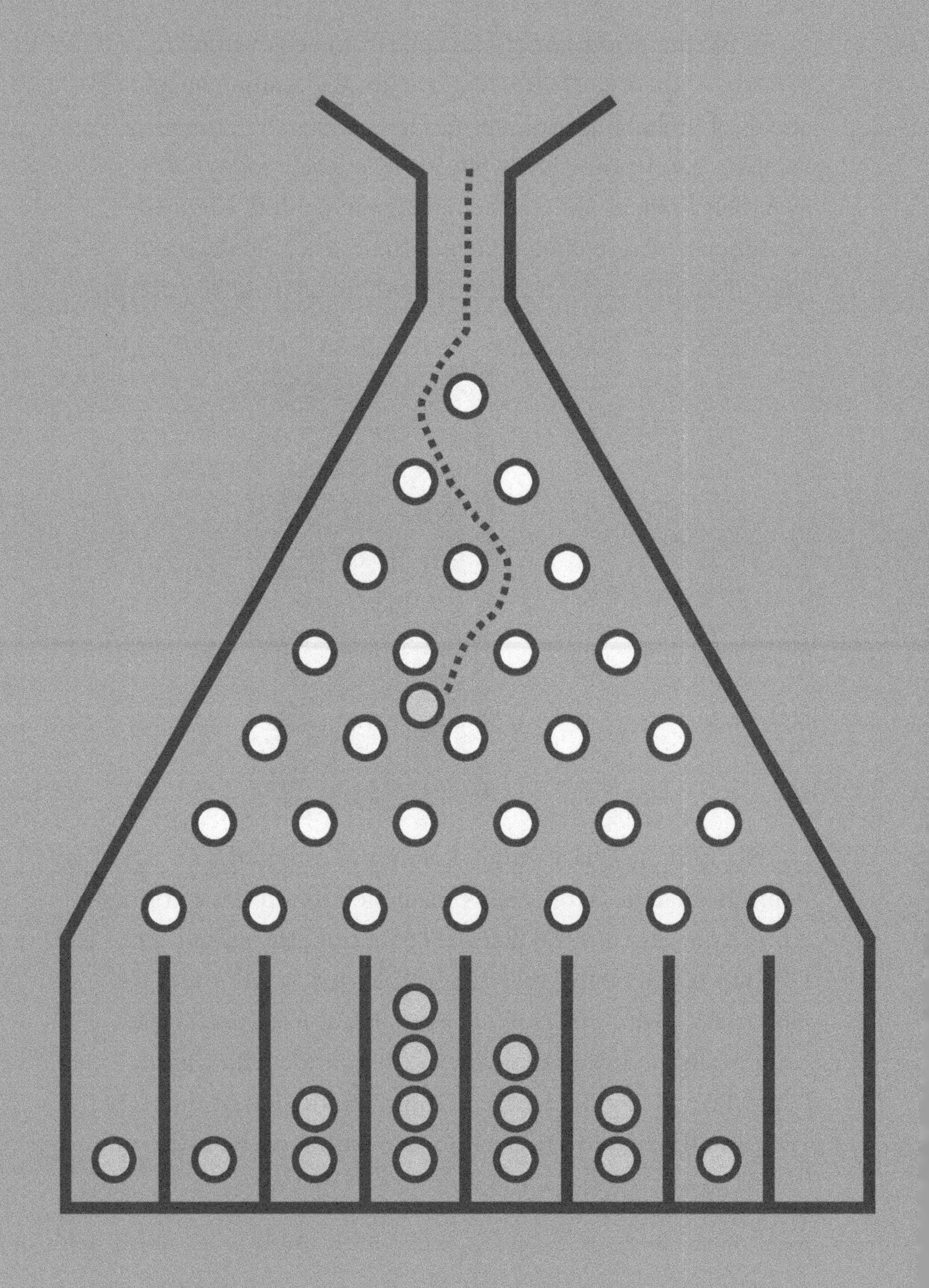

Since balls have an equal chance of going in either direction, one might think that it would be impossible to predict what the outcome will be after the balls are released to flow through the board. However, it's not only possible to make a prediction – it is virtually certain that no matter how many times you repeat the experiment, you will produce roughly the same outcome:

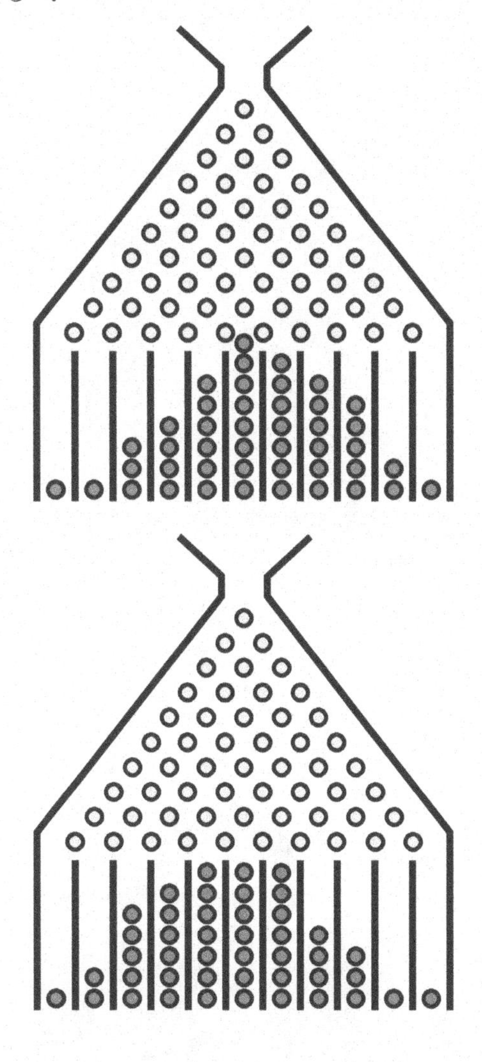

What is going on? How can such an inherently random process produce such consistent results? By the way, it is genuinely random – watching a single ball pass through the board really tells you absolutely nothing about what path the next ball will take; each ball is completely independent from the ones that came before and it is free (insomuch as inanimate objects are free) to take absolutely any path to get to the bottom. The key to understanding the behaviour of the quincunx is not to think about each ball individually, but to view the whole group of balls as a single entity that is governed by certain rules.

Let me explain.

To help us visualise what's happening and understand the pattern that makes the quincunx do its thing, let's consider a much smaller version with far fewer pins. This will shrink the problem down to a scale we can manage. If the quincunx had four rows of pins, then there are exactly 16 possible paths for a ball to take to get to the bottom. Here's what each one looks like:

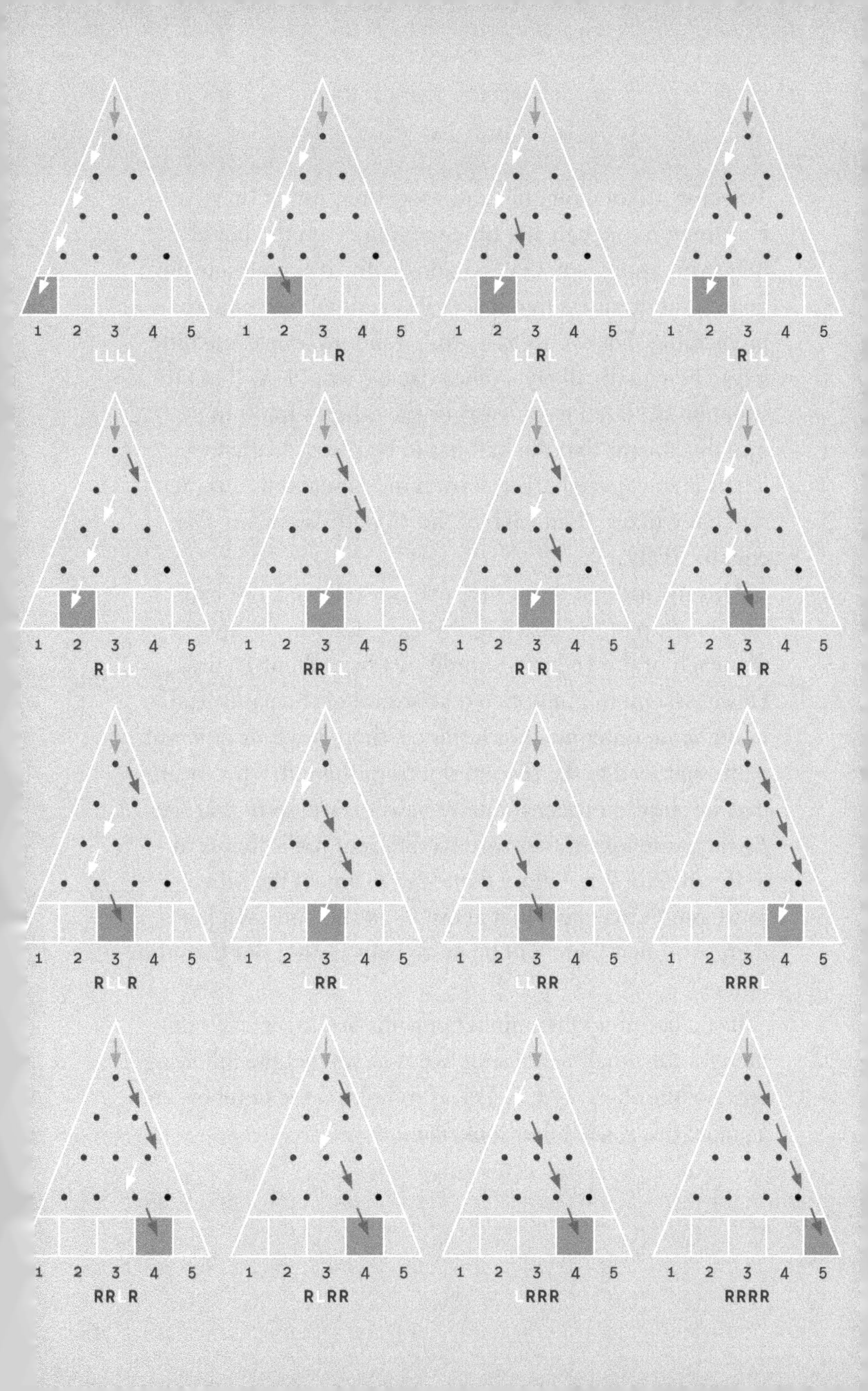
1 2 3 4 5
LLLL
1 2 3 4 5
LLLR
1 2 3 4 5
LLRL
1 2 3 4 5
LRLL
1 2 3 4 5
RLLL
1 2 3 4 5
RRLL
1 2 3 4 5
RLRL
1 2 3 4 5
LRLR
1 2 3 4 5
RLLR
1 2 3 4 5
LRRL
1 2 3 4 5
LLRR
1 2 3 4 5
RRRL
1 2 3 4 5
RRLR
1 2 3 4 5
RLRR
1 2 3 4 5
LRRR
1 2 3 4 5
RRRR

Seeing all the potential paths together, rather than focusing on one ball at a time moving down the board, helps you appreciate that, paradoxically, it is precisely the board's randomness that makes its eventual outcome so predictable. For events to be truly random, every outcome has to be equally likely – otherwise we would say that the situation is biased in one way or the other. From pin to pin, this means that the ball has to be just as likely to go left as it is to go right. But if each individual left or right is equally likely, then each of the 16 paths shown is also equally likely.

This means that if, for instance, we released 160 balls from the top of the board in this case, we would expect that each of the 16 paths should be taken about 10 times. However, you might notice that several of the paths lead to the same outcome. For instance, there are four different paths that lead to the second slot from the left. This means that we should expect about 40 balls to end up in that slot. By the same token, there are six different paths that lead to the middle slot – more than lead to any of the others – so we ought to expect that about 60 of the balls will land there, and that there will be more balls in that slot than the others.

If you count up the number of paths leading to each slot, then for this small board with five slots you get the following series of numbers: 1, 4, 6, 4, 1. If you plot these numbers on a graph, this is what they look like:

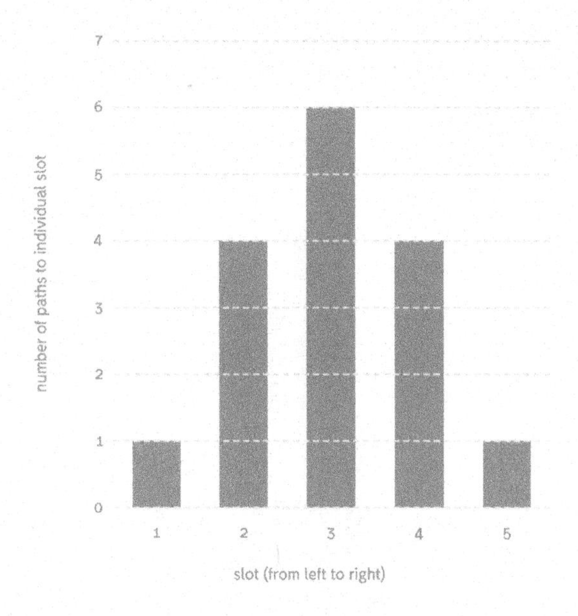

We can repeat this exercise with an imaginary quincunx that is taller and has more slots – say, nine instead of five. If we count the number of paths again, you get these numbers: 1, 8, 28, 56, 70, 56, 28, 8, 1. This time, the graph looks like this:

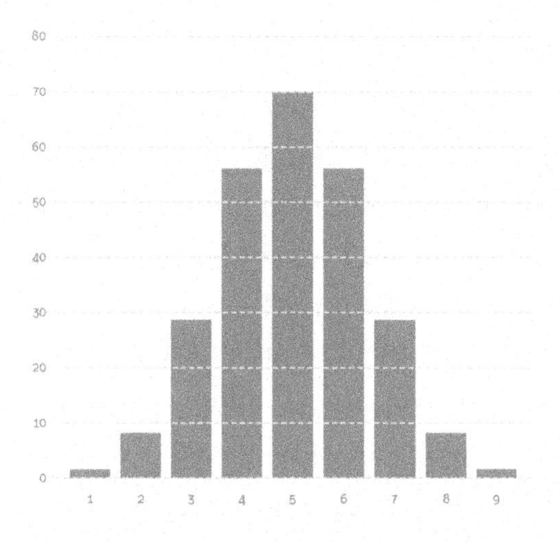

And if we increase the number to 21 slots, then this is the shape that emerges:

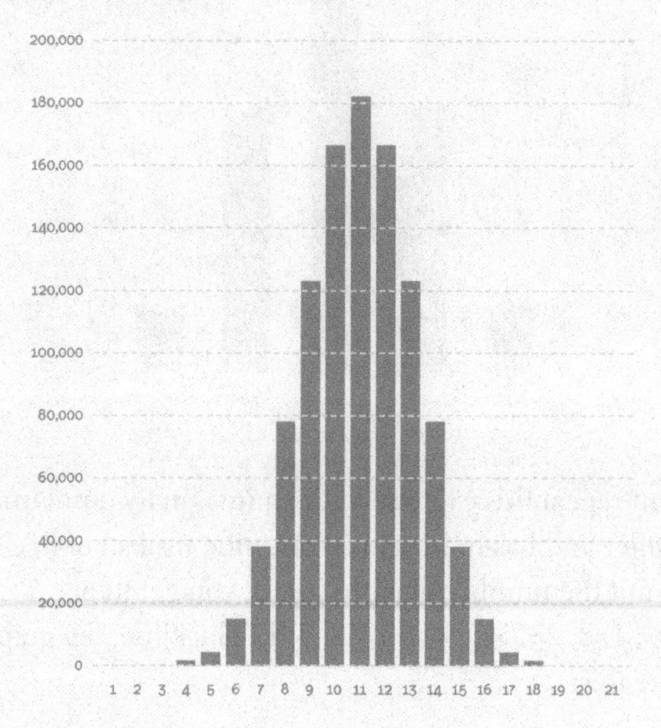

Look familiar?

As we noticed back in our chapter about fractals ('Lightning through your veins'), poetry is the art of calling the same thing by different names: of finding new and imaginative ways to stretch the limits of language and express old and universal realities. But mathematics hones our ability to look at things that appear to be completely different on the outside, but are unified by a single and beautiful pattern. We can call different things by the same name in mathematics because we see the thread that runs through them all.

Many people would call the shape we see in the preceding graphs a bell curve, but mathematicians call it 'the normal distribution'.

This reflects the fact that all groups influenced by randomness look like this. In other words, when scores are distributed with an element of randomness, this is what they normally look like.

Whether the group is of people, or test scores, or events (like balls travelling down a quincunx), they all share this characteristic shape.

Henri Poincaré, a Frenchman who lived from 1854 to 1912, reportedly understood this fact about the normal distribution. While he was most known for his contributions to mathematics and physics, an old anecdote relates the way that he used his statistical smarts to catch an underhanded baker.

The story begins with Poincaré's suspicion that his local bakery was cheating its faithful customers by intentionally selling them loaves of bread that were lighter than advertised. They were supposed to weigh exactly one kilogram each – and being precise with weights and measures was something that the French were supposed to take pride in, since France was the home of the International Bureau of Weights and Measures. The Bureau was charged with stewarding the original International Prototype of the Kilogram, from which the kilogram was officially defined around the world – so weight was something they should definitely have had the ability to get right!

Poincaré decided to use statistics to get to the bottom of things. Every day for an entire year he dutifully purchased a loaf of bread, brought it directly home and weighed it. By year's end, he had an impressive collection of data to draw on – and he used it to plot a distribution of his measurements over the months. He showed that it fit a bell curve with a mean of 950 grams and a standard deviation of 50 grams. This meant that, according to the way the normal distribution spreads itself out, only 16% of the baker's loaves weighed at least 1000 grams, while the other 84% were lighter than advertised. He brought this to the attention of the authorities, who issued the baker with a warning.

The following year, Poincaré (ever the sceptic) decided to continue the practice of weighing his bread each day. Initially he was delighted, because the average weight seemed to be coming out at exactly 1000 grams – just as it should be. But as the year wore on, he again started to become suspicious. At the end of the year, he again brought his findings to the authorities, who promptly issued a fine. What had gone wrong?

Poincaré had noticed that the shape of the distribution was not even – it was asymmetric, with more weights found on the heavier end of the scale. Procedures subject to ordinary random factors, like those of the quincunx, cannot help but produce the telltale symmetry of the normal distribution. Poincaré's conclusion was that the loaves he'd purchased in the second year were not subject to random factors, but had been selected on purpose. The baker was still making 950-gram loaves – but he carefully chose the heavier ones to give to Poincaré after he had made his first complaint.

The principles that lead to the normal distribution are at work beneath many everyday phenomena. This is what allows us to effectively predict the future without realising it. How long will it take for you to arrive at your destination if you are catching a bus from your home into the city? It's theoretically possible to determine the exact distance covered along the roads between A and B, and then divide by the speed limit of each road to give a time. But this is a fool's errand: there are traffic lights that will cause you to stop along the way, interrupting your journey an uncertain number of times. The speed limit of a road is likely to be a poor indicator of the actual speed you will travel on it, especially during peak hour, since the volume of traffic might turn your highway into little more than a car park for half of the journey.

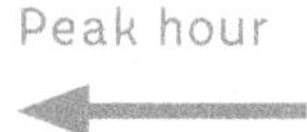

The normal distribution can rescue us here. While each individual journey is difficult to predict – just like the path of a single ball down the quincunx – the group of all journeys from your suburb into the city cannot help but settle itself inexorably into something resembling the normal distribution. The traffic lights on the journey let some vehicles through and stop others, like the pins sending some balls left and some right. And just as it is highly unlikely to go left at every single pin in one journey, it's unlikely that you will be stopped at every light (or, conversely, that you will not be stopped by any of them). As we take more and more journeys, the times add up like balls at the bottom of the quincunx. This is how online maps build what's called a probabilistic model for predicting how long a typical journey will last, which is how they can give us estimated times of arrival that are sometimes stunningly accurate.

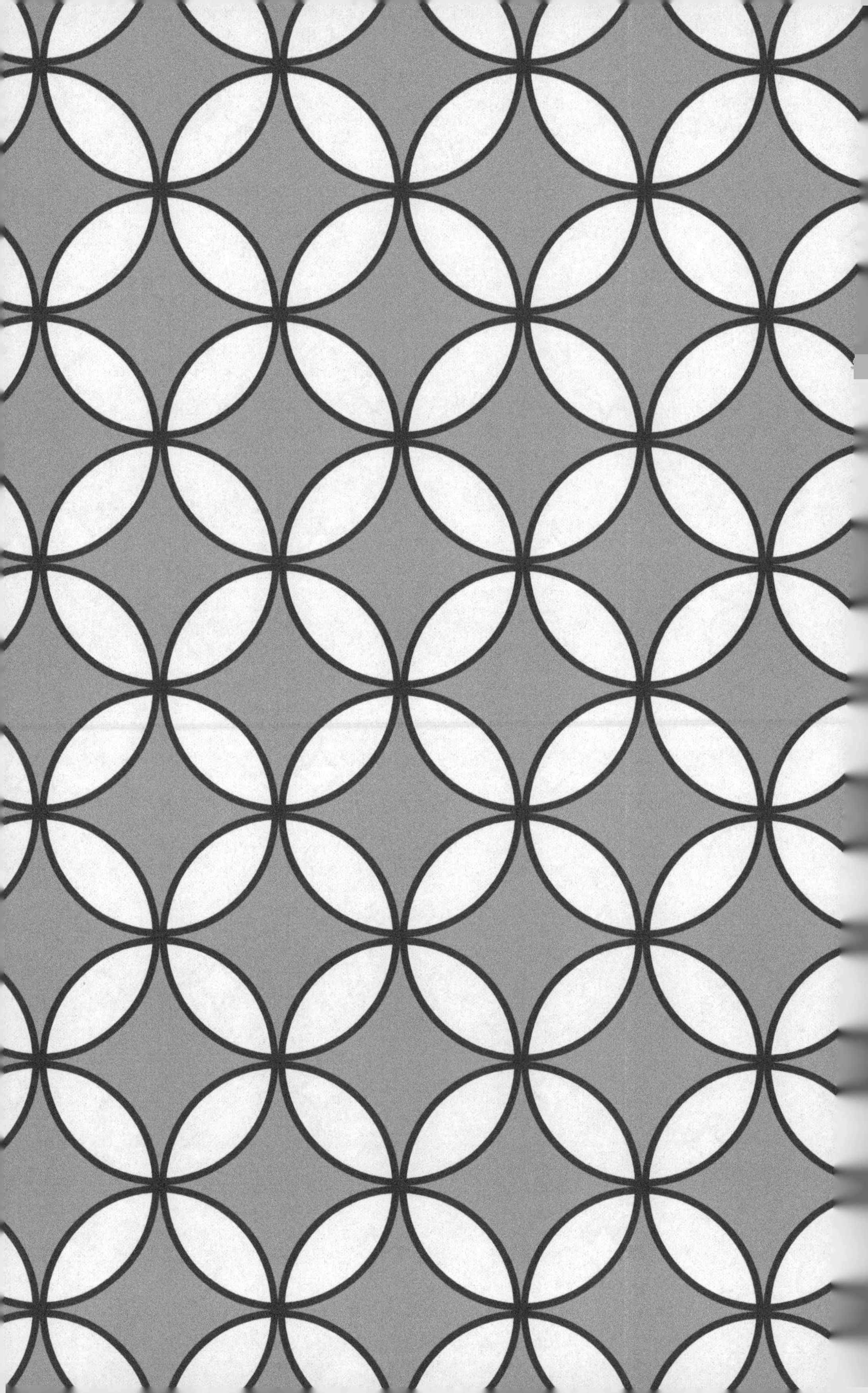

CHAPTER 13

KILLER BUTTERFLIES

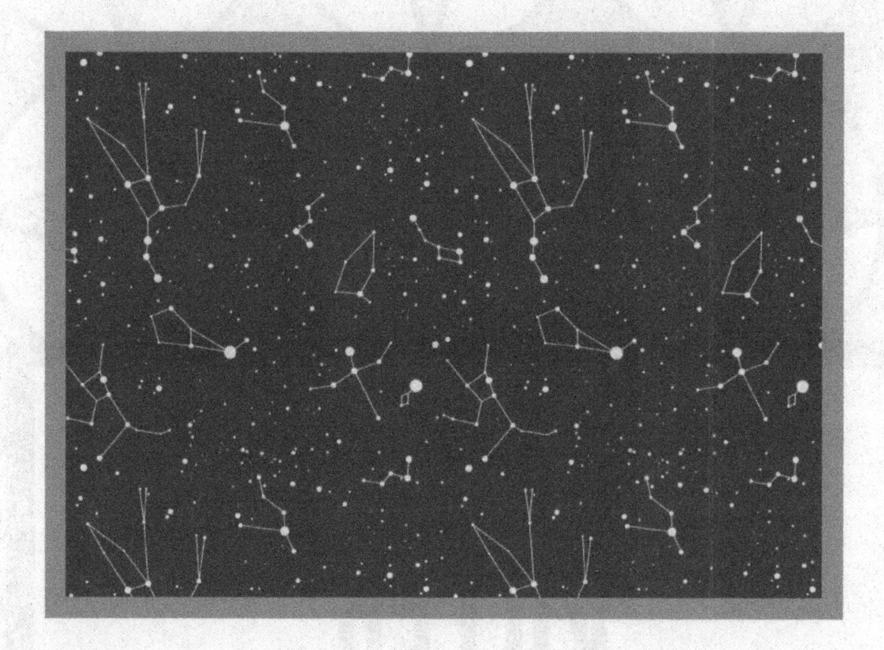

In the previous chapter, we looked at how apparently random processes are often surprisingly predictable. Not everything goes so smoothly when you're trying to see into the future, though. Some things, such as the weather forecast, are famously unreliable for anything more than a day or two in advance. Our inability to accurately foresee the weather is even more perplexing when you remember that weather – being something that is experienced universally by people in all places at all times throughout history – has been the subject of intense study down the centuries, and so you would think its patterns should be incredibly well understood by now. So how has weather been able to defy our predictive powers for so long? Again, mathematics can explain – and this time, the key is in a branch of mathematics called 'chaos theory'.

In common language, 'chaos' means 'a state of disorder'. This is as opposed to 'cosmos', which refers to something orderly – a throwback to the early pioneers of astronomy who saw the rhythmic motion of celestial objects as a pointer to divine design. It's easy to contrast chaos with cosmos: to visualise cosmos, picture the absolutely predictable shapes and arrangement of the constellations in the night sky. It was the very predictability of the stars that was so useful to ancient mariners: if you could learn to read their positions and compare them to known star charts, you could accurately work out your position in the middle of a featureless ocean.

Conversely, when we picture chaos, we might think of a sandpit that has been recently visited by an unruly toddler, who has come along and mercilessly crushed all the beautiful constructions made by the children who left five minutes earlier. No one can predict the exact position of every grain of sand in the pit once that toddler is finished with it, because her actions are not governed by any orderly principles or rules (except, perhaps, to destroy everything in sight).

Something like the weather certainly seems to be more like the sandpit than the stars. While the broad brushstrokes of the seasons and their undulating temperatures are easy to keep track of, the weather of any individual day might as well have been decided by our irascible three-year-old. As our understanding of the physical universe has grown, though, it's become increasingly clear that things sometimes only have the appearance of disorder

and randomness. Many things are actually completely predictable, as long as you know all the information about them in advance.

For instance, a coin flip is often used as the quintessential emblem of blind chance. But if you knew everything about the coin and how it is launched into the air – its weight, the force with which it is flipped, the humidity of the air and so on – it would truly be possible to know the outcome of the coin flip before it happened. Mathematicians call this a 'deterministic system' – the state of things at a certain point in time is completely determined by the events leading up to the event in question, with no genuinely random things ever actually occurring.

The mathematician's definition for chaos includes such situations: a system – such as the roll of a die or the changing conditions of the weather – can exhibit 'chaotic' behaviour, even if there is nothing truly random within it. It's worth pausing to appreciate the craziness of this idea. It's as if you discovered that the toddler had actually gone into the sandpit with a perfectly detailed blueprint in her mind for where she wanted all of those millions of grains of sand to go, and she carried out the precise set of motions in the pit to produce that exact result. How could this be?

To understand, we need to meet a very important mathematical object called a map.

Mathematical maps are quite different from their cartographical cousins, but they share the same intent: to show you the relationships between different objects.

Train maps show you the relationships between stations (in terms of connections within the network) and street maps show you the relationships between geographical locations (in terms of distance and direction). Mathematical maps show you the relationships between numbers, in terms of how one number can lead to another once certain mathematical operations have been carried out on it.

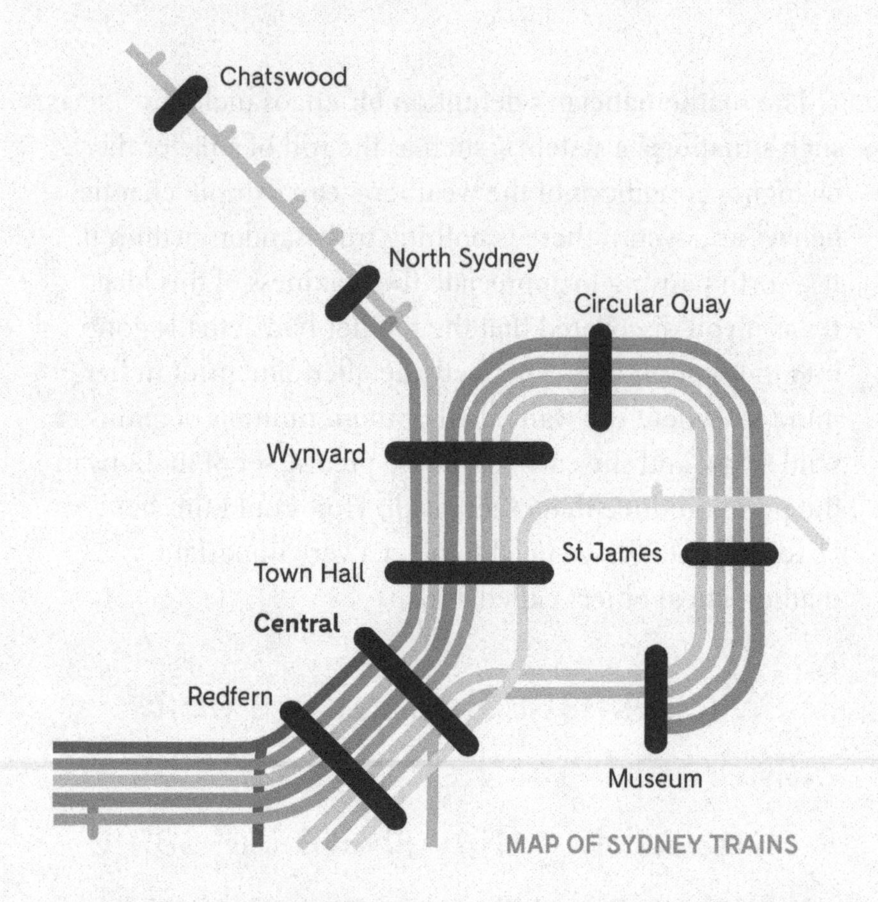

MAP OF SYDNEY TRAINS

It's worth noting that maps we deal with in the real world often work according to different rules, which we tacitly acknowledge but seldom ever think about. For instance, street maps almost universally come with a scale: a long distance on the map corresponds to a long distance in reality. Train maps don't share this rule: two stations may appear very close to each other on the map, but are actually separated by a huge distance on the train line. In the same way, different mathematical maps have different rules for how numbers relate within them.

Let me show you what I mean. Here's an example of a mathematical map:

$$X_n \times 2 = X_{n+1}$$

All this means is that you start with any number you like, and to get to the next one you multiply what you've got by two. We simply repeat this step over and over again – a process called 'iteration' – to move forward through the map. So if I started with the number three, this is the result I would get from my mathematical map:

Step 1	Step 2	Step 3	Step 4	Step 5	Step 6	Step 7
3	6	12	24	48	96	192

Since you're doubling the numbers from one step to the next, we could call this the doubling map. I can start with a different number if I want, so here's what happens if I start from four:

Step 1	Step 2	Step 3	Step 4	Step 5	Step 6	Step 7
4	8	16	32	64	128	256

And here's what happens if I start from some numbers in between three and four:

Step 1	Step 2	Step 3	Step 4	Step 5	Step 6	Step 7
3.1	6.2	12.4	24.8	49.6	99.2	198.4
3.5	7	14	28	56	112	224
3.9	7.8	15.6	31.2	62.4	124.8	249.6

What I want you to notice about this set of values is that, perhaps exactly as you'd expect, the doubling map is highly predictable. If you increase the starting number by a small amount, then the final result is bigger by a small amount. If you increase the starting number by a larger amount, the final result grows accordingly. This means that if you were to arrange the rows in ascending order based on their starting number, then they would also be in ascending order based on their final number:

Step 1	...in-between steps...	Step 7
3		192
3.1		198.4
3.5		224
3.9		249.6
4		256

Using this knowledge, if I gave you a starting number and asked you to guess the final number, it would be completely plausible for you to make a very good estimate. For instance, where would we end if we started with 5? How about 6? Try thinking about the value of the final number if you started with something even bigger, like 10. Sharp readers will have worked out by now that you can go straight from the first number to the final number if you just multiply by 64. As a result, 10 will become 640.

What this shows is that in this situation, knowing the starting number – which mathematicians often call the 'initial conditions' – allows you to easily predict the final number with great confidence. The rules of this map aren't too complicated, so maybe that comes as no surprise to you.

Now I want you to have a look at the following set of starting numbers and final numbers, which have resulted from a different mathematical map:

Starting number	...in-between steps...	Final number (approximate)
0.0001		0.243
0.0002		0.880
0.0003		0.477
0.0004		0.174
0.0005		0.604

The more you look at the numbers coming out of this map, the more confusing it appears to be. The differences between the starting numbers is tiny: only one ten-thousandth from one row to the next. By contrast, the differences between the final numbers are enormous: a small change in the initial conditions results in a change thousands of times larger once our numbers have made their way through the map. In addition, the order of the starting numbers tells you nothing about the order of the final numbers: sometimes increasing the starting number leads to an increase in the final number, but the size of the increase appears to be random – and just as often there is a decrease rather than an increase. What on earth is going on here?

These numbers result from a particular map called the logistic map. You might expect the logistic map to be a garbled mess of complicated symbols and arcane mathematical expressions, but in fact it is an incredibly concise object:

$$4x_n \times (1 - x_n) = x_{n+1}$$

The logistic map's ability to generate seemingly chaotic behaviour is even more striking when you pull back the curtain a little to reveal more of the steps (rather than focusing on the starting numbers and final numbers alone). Here's a quick snapshot of the in-between steps that I didn't show in the table:

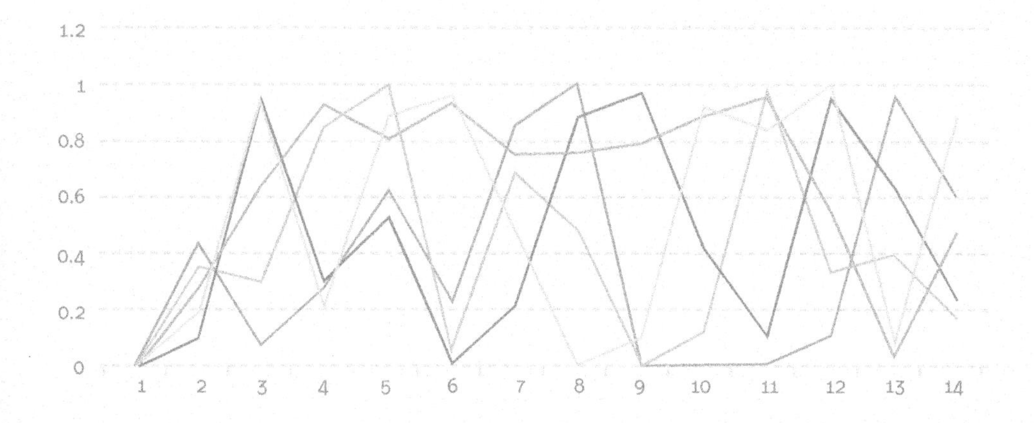

As you follow a jagged line from left to right, you are watching the logistic map transform a number by taking it and reinserting it back into its algebraic set of rules. The lines are shaded differently to make it a little easier to track where they came from originally: smaller numbers are represented by darker lines and larger numbers are represented by lighter lines. You can see that all the lines seem to originate from a single point on the diagram because the differences between the starting numbers (as shown in the table) are so miniscule.

The logistic map is an archetypal example of mathematical chaos. This is because it exhibits the crucial condition of chaos: an extreme sensitivity to initial conditions. You only have to change the starting number by a tiny amount, and the line tracking its journey gyrates wildly in response. In the diagram on the next page you can see the continued journey of 0.0001 (dark) and 0.0002 (light) as they are twisted and turned by the logistic map.

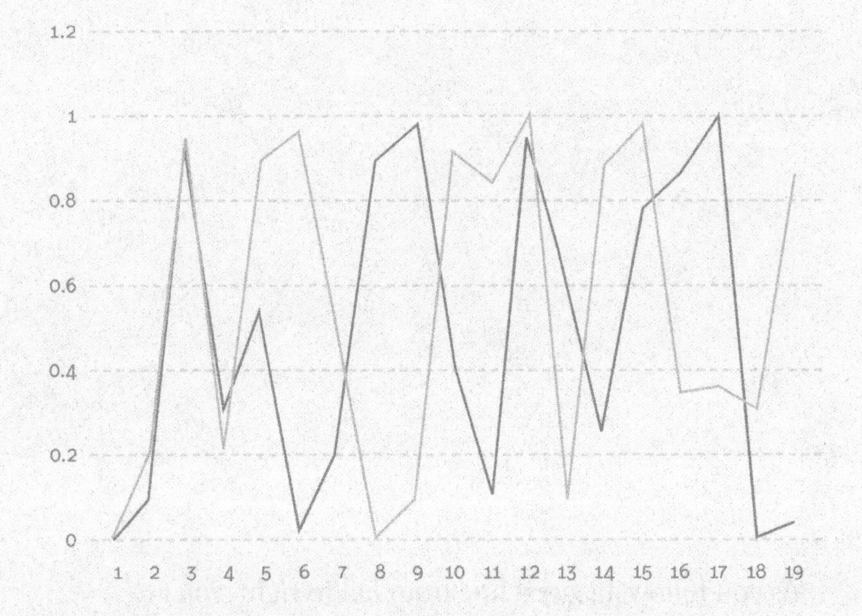

'Sensitivity to initial conditions' has been nicknamed the 'butterfly effect', in reference to the idea that a butterfly flapping its wings (a tiny initial change) can lead to a cyclone on the other side of the world due to the way it has imperceptibly changed the flow of air in the atmosphere. While it may seem crazy to think that such a small creature can bring about such changes on such an enormous scale, this is exactly the reason that our weather forecasts struggle so much to predict things in advance.

Imperfect weather reports aren't due to our inability to track rogue winged insects so much as they are a natural consequence of our limited measurement instruments. Meteorological bureaus around the world collect data about the current weather state – barometric pressure, temperature, humidity, wind speed, nearby ocean currents and the like – and insert these pieces of data into a mathematical map which acts like a model to predict these same pieces of information for the days ahead. But no matter how accurate the model may be, the pieces of data will always have a flaw: their limited potential for accuracy. We may know the temperature to within 0.0001 °C, but as was demonstrated by the logistic map, a difference of just 0.0001 can still lead to an enormous change just a few steps later.

In fact, the graph of 0.0001 and 0.0002 gives a great sense of how this change actually unfolds. Suppose our thermometer measures the temperature as 25.0001 °C when in reality it is 25.0002 °C. The dark line represents our prediction of how the temperature should fluctuate while the light line represents how it will actually progress. If you imagine each step from left to right along the map as a day, you can see that the first three days are never identical, but they look remarkably similar. They only differ by small amounts, which would mean that the weather report is reasonably accurate. But at day four, all bets are off. The tiny changes give way to significant differences, and before long the temperature maps appear to have almost nothing to do with each other.

CHAPTER 14

ILLUMINATI CONFIRMED

In the chapter 'Precognition for dummies', we explored an idea called the normal distribution by looking at a wonderful little device called a quincunx. By thinking about imaginary boards of various sizes, we realised that we could accurately predict the final state of any board – even enormous ones with hundreds or thousands of rows and slots – despite the fact that the path of any individual ball through the board is genuinely random. Our prediction was powered by the fact that some simple counting could work out the total number of paths to each slot at the bottom of the board. The basic fact of randomness – that all paths are equally likely – does the rest of the work.

I want to return to that moment and draw your attention to an astonishing object that emerges from this little exploration. We determined that if a quincunx had five slots at the bottom, then from left to right, the slots had the following number of paths leading to them: 1, 4, 6, 4, 1. If a board had nine slots instead, then the number of paths (again, from left to right) would be: 1, 8, 28, 56, 70, 56, 28, 8, 1.

If you take every sequence of numbers that is made in this way, they can be arranged into a triangle like this:

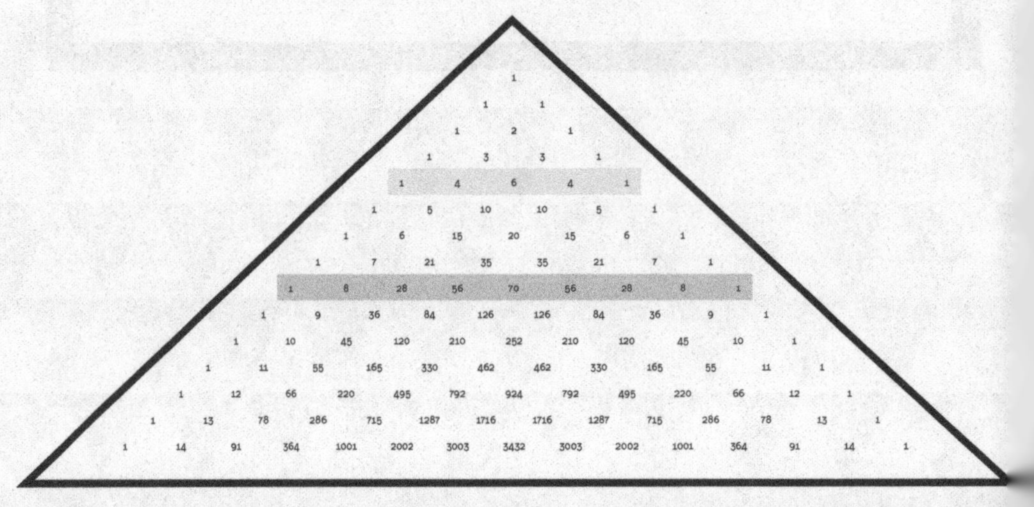

Notice the fifth and ninth rows from the top, which correspond to the numbers for our quincunxes with five and nine rows respectively. This perplexing pyramid of numerals has been discovered and rediscovered throughout history and in a variety of different cultures, so it has been called many things: Yang Hui's Triangle, the Staircase of Mount Meru, Tartaglia's Triangle and Khayyam Triangle, among others. However, if you're reading this book in English, you most likely know it by the name Pascal's Triangle.

Pascal's Triangle can be understood completely without a quincunx. You start with a 1 at the top of the triangle, then you find every number below by adding the two numbers above it. (If you are on the left and right 'edges' of the triangle, you imagine a zero on the outside.) For instance, the ninth row (highlighted in the triangle opposite) can be calculated using numbers from the eighth row:

$$7 + 21 = 28,\ 21 + 35 = 56, \text{ and so on.}$$

This is the first reason that mathematicians have stumbled upon Pascal's Triangle again and again: it's accessible. All it takes is addition, so even a small child in primary school can (with time and patience!) calculate many of the rows by hand. But lots of things are easy to work out yet worthless! The second reason that this triangle has enduring beauty is because it's like a rich diamond mine, replete with treasures that are almost as easy to find as the initial construction of the shape. For instance, have a look at what happens if you highlight all the even numbers you can find. The following pattern emerges:

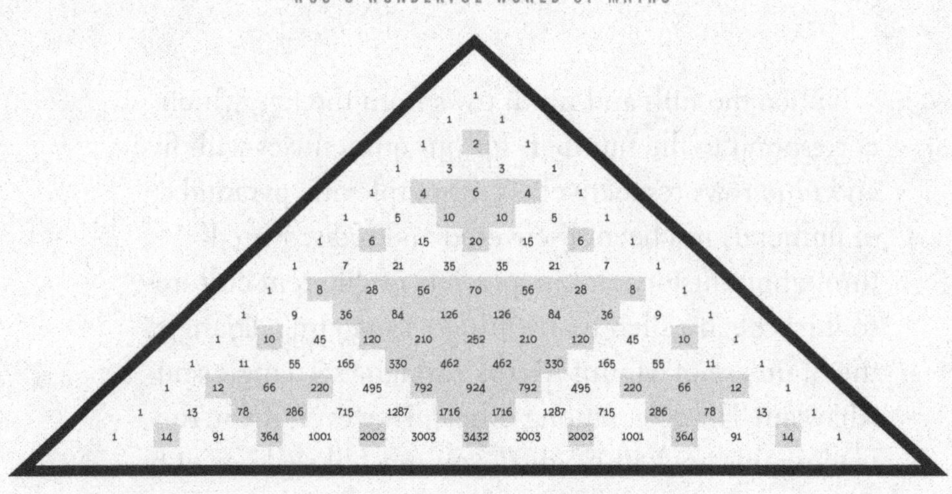

Even numbers aren't the only ones that give an interesting pattern. This is what happens if you highlight all the multiples of three:

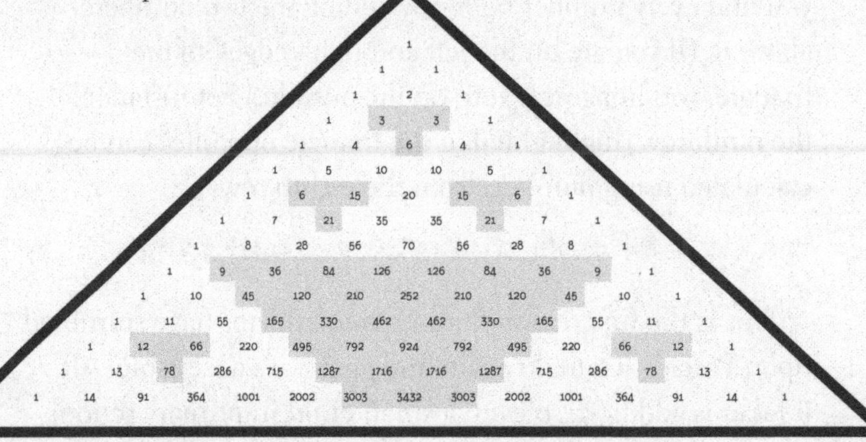

In fact, multiples of *every number* give interesting versions of this pattern. See the multiples of five in the first triangle on the next page.

Just like Pascal's Triangle, this pattern of triangles also has its own special name: it's called Sierpinski's Triangle, and it's a fractal with self-similarity just like the lightning bolts and blood vessels we looked at earlier.

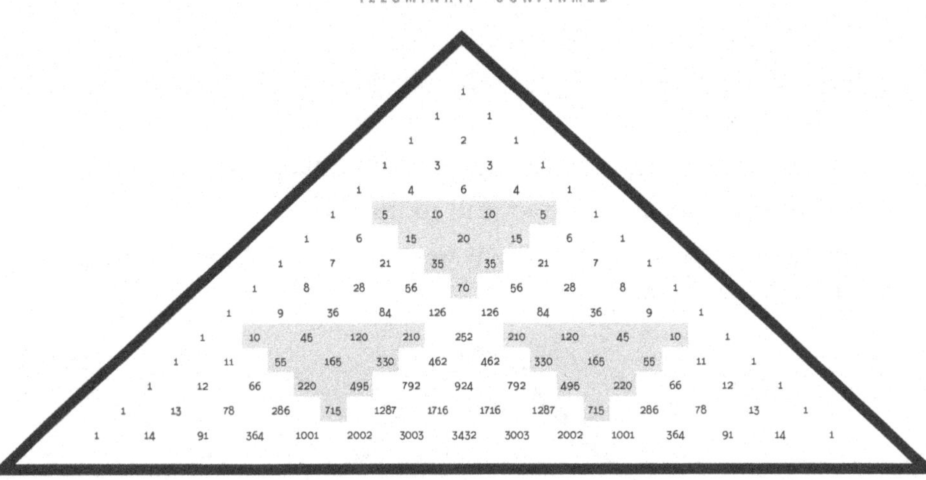

There are even more stunning patterns if you're willing to keep digging. For instance, if you add up the numbers horizontally to find the sum of each row, then from the top you get: 1, 2, 4, 8, 16, 32 and so on. Adding the rows gives powers of 2, each row having double the total of the row above it.

If you think the rows are weird, we're just getting started. For example, have a look at this diagonal of numbers:

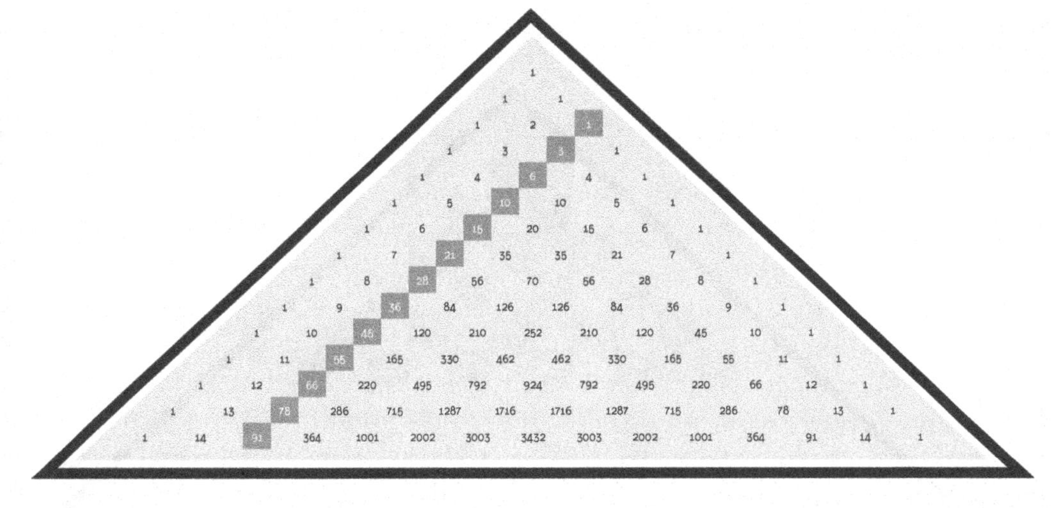

These pearls are the 'triangular numbers': 1 is the first triangular number, 3 is the second triangular number, 6 is the third triangular number and so on. The easiest way to see why they have that name is to step back and look again at Pascal's Triangle:

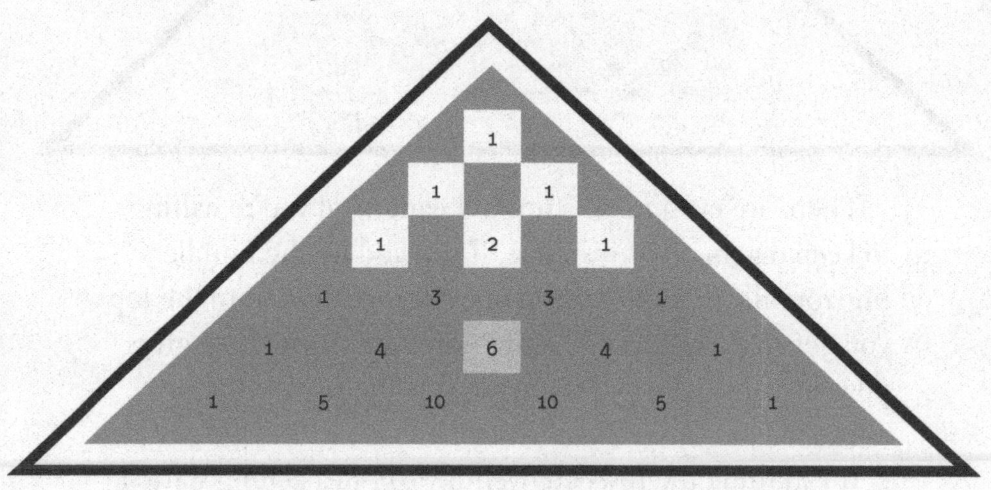

The number 6 is the *third* triangular number; there are 6 numbers in the first *three* rows of Pascal's Triangle.

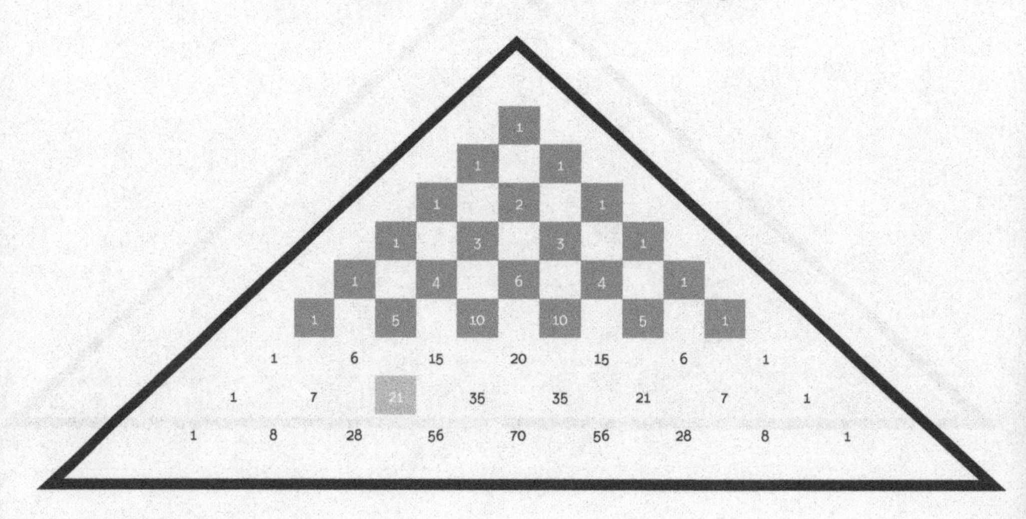

The number 21 is the *sixth* triangular number; there are 21 numbers in the first *six* rows of Pascal's Triangle.

Pascal's Triangle also has a curious relationship with the primes, which we first met in the chapter 'Unbreakable locks'. The relationship becomes visible if you ignore the 1s on the edges, and then pay attention to the rows that start with a prime number.

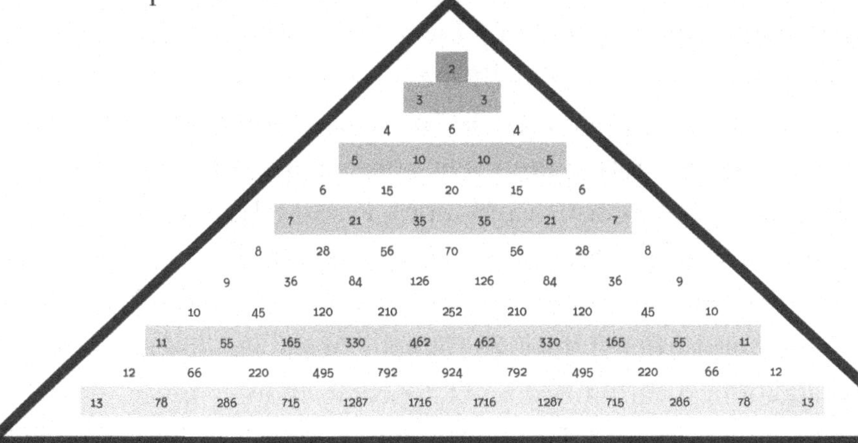

It might not be apparent on first view, but the numbers on each row share a remarkable property: they are all multiples of the first number in the row. See:

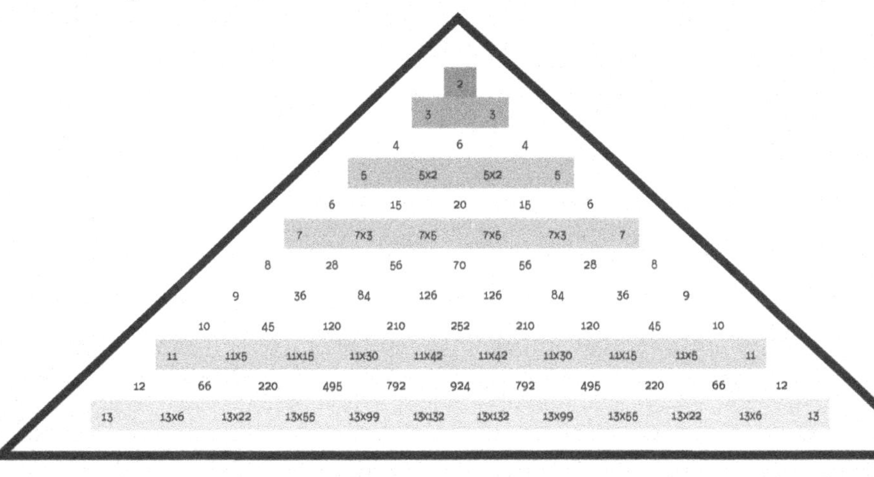

Why should anyone care about this?

Well, think of it this way. Pascal's Triangle is the mathematical equivalent of the Cullinan Diamond. Weighing 3,106.75 carats, it was the largest gem-quality rough diamond ever found. One of the most astonishing things about gems, especially the Cullinan, is that they are created from simple materials (in the case of the Cullinan, carbon atoms – lots of them!) by completely natural processes (heat, pressure and time). Pascal's Triangle consists entirely of the simplest kind of number – the counting numbers, which we will explore in Chapter 16 – and is constructed through the fairly mindless process of addition. It certainly doesn't appear as though any deliberate or careful design went into its construction. Yet, just like the Cullinan, it brims with a beauty that an artist or architect could only dream of. Turning it in your hand just a tiny amount and looking at it from a new angle reveals new patterns and colours. The simplicity and abundance of patterns in Pascal's Triangle is ready and waiting to be discovered by anyone willing to play with numbers.

CHAPTER 15

TWINKLE, TWINKLE, LITTLE STAR

Quick, picture a star shape in your mind! What does it look like?

There's a good chance that you pictured a shape with sharp points directed outwards and with straight edges between each of its corners. Something like this:

We teach our babies and toddlers that this is a star shape, and it doesn't take long for the indoctrination to stick. The obligatory solar system diorama that all children seem to complete in primary school is littered with 'stars' that invariably look like this:

The conspiracy continues well into adulthood. For instance, we call this sea creature a starfish:

And all around the world, flags bear 'stars' that all resemble this familiar pointed shape. The US flag is even referred to as the 'Star-Spangled Banner' for its prominent use of this figure:

But it's all a bald-faced lie.

There isn't a single star in the universe that is pointy like this – look, I'll even prove it to you with photographic evidence!

Despite their name, stars aren't star-shaped.

All stars are, in fact, spherical. This leads directly from the fact that gravity attracts objects (in this case, the white-hot plasma that makes up most of a star's mass) together in proportion to their distance from each other. At a certain distance, the star's gravity isn't strong enough to hold on to stellar material, and so that marks the size of the star. In two dimensions – like on a flat piece of paper – the shape you trace out when you measure a fixed distance from a central point is a circle. In three dimensions – like space as we know it – you get a sphere.

This universal piece of astronomic geometry seems to have gone by mostly unnoticed by civilisations throughout history. Why is it that humanity has so consistently represented something so quintessentially round with its geometric opposite, a pointy and prickly imposter?

Before we answer this question, it's worth spending a few moments to appreciate just how much geometry – the study of shapes – has been caught up with our understanding of the stars. The celestial bodies we can see in the night sky have captured the imagination and intellect of people the world over, and every culture has ascribed some kind of meaning to the stars. Whether it's been through the inventive myths of astrology or the more modern approaches of scientific rationalism, we've always agreed that the stars are hugely important and so it makes sense that stargazers through the ages have sought to understand them mathematically.

For instance, it's fairly common knowledge that if you are measuring an angle then we say there are 360 degrees in a whole revolution. It's why the phrase 'doing a 180' has entered our language to mean a complete reversal in your decision or approach. But have you ever wondered why it is 360? Who decided that and why? It's so deeply written into our thinking that many of us have trouble imagining that there could be any other way. But there *are* other ways – for instance, in some parts of Europe they use the 'gradian', of which there are 400 (not 360) in a full revolution – much nicer to handle with our decimal system. A quarter turn, also known as a right angle, is exactly 100 gradians – which makes a whole lot more sense than the comparatively arbitrary 90 degrees that we're used to.

People have divided up a full revolution into 360 degrees since ancient times, so the exact reasons why 360 was chosen are well and truly lost to history. But there are two suggestions that make so much sense in the context that it's hard to imagine they were not each involved in some way.

The first reason has to do with the factors of a number. When we think about whole numbers, factors can be thought of as the ways you can divide up a number into equal groups so that there are none left over. The number 10, for instance, has four factors: 1, 2, 5 and 10. That means we could divide 10 into 1 group of ten, or 2 groups of five each, or 5 groups of two each, or 10 groups of one each.

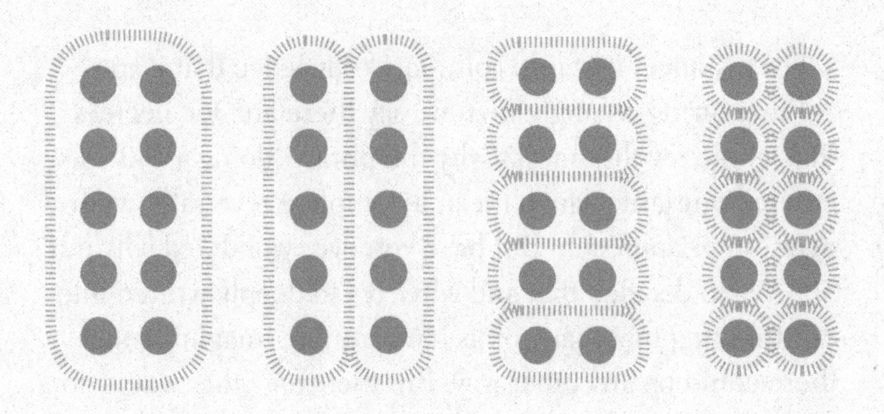

It's nice when a number has lots of factors, because that means you can divide it up easily into many different even sections. In the context of angles and rotation, that means we can neatly describe several kinds of partial turn with whole numbers. For instance, a turn that goes halfway around a circle would be 180 degrees while a turn that goes a third the way around would be 120 degrees. The decimal system hates being divided into thirds because 3 isn't a factor of 10: a turn that goes a third the way around a circle would be an awkward 133.333333 . . . gradians.

Just how many factors does 360 have? A lot! In fact, every number from 1 to 10 – except for 7 – is a factor of 360 (and 360 is the smallest number in existence to have this quality). If you keep on going, you'll find 360 has exactly 24 factors, which is not bad considering the previous number (359) has only 2 factors and the next number (361) has only 3 factors!

So, there's the first reason for 360: its special ability to be divided up cleanly into so many different chunks makes it very useful for practical purposes. But I did mention that there was a second reason, and this is where we see 360's connection to the stars. For centuries, humanity has navigated its way across the oceans using the stars. Without alternative reliable physical landmarks to guide them, sailors were forced to look at the only constant feature – the sky above them – to determine where they were.

But celestial navigators rapidly discovered a problem with using the stars to find their location. The stars aren't fixed – certainly not from our perspective. Even a cursory observation with the naked eye will reveal this fact, no telescope necessary – as long as you have the patience to wait a few hours.

The stars appear to move, because the Earth, from which we observe the stars, is itself not still. For starters, the earth is rotating on its axis like a child's spinning top. As a consequence, our view of the stars is like the view from a merry-go-round: the fair may be motionless around you, but it appears to move because our perspective on it keeps changing. If you point a camera at the night sky and leave its lens open long enough, you will see the circular path traced out by the stars as they appear to move.

It's really us that's doing the moving, pivoting around and constantly adjusting our point of view.

If you take a photo of the stars from a fixed position, and then take another photo exactly an hour later from the same spot, you will find the stars in different positions. But if you wait 24 hours, the stars will have 'returned' to the spots you noticed them in the day before.

Well, almost. That's because there's another kind of way that the Earth moves. Unrelated to the fact that it spins, our planet also orbits the sun. This means that even if you took your photo of the night sky at exactly the same time each day – say, midnight – the Earth's orbit would actually face you at a different patch of the night sky each time!

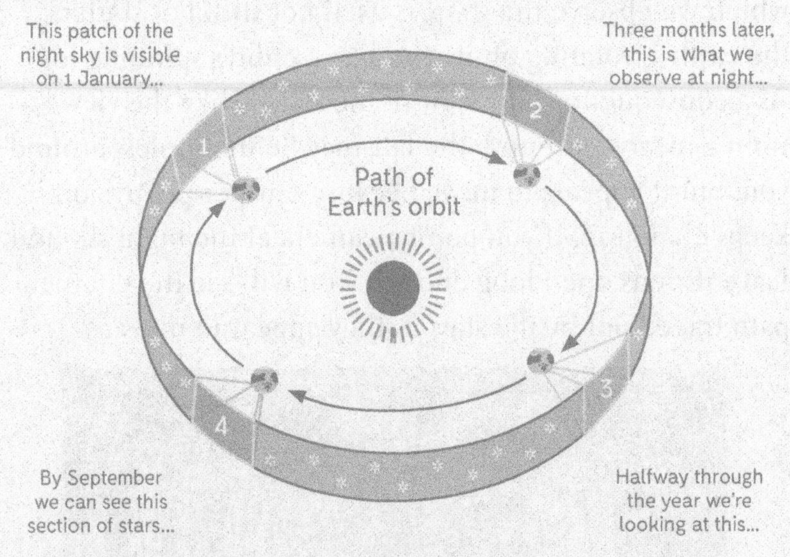

However, after a certain amount of time, you'll find yourself back in the position you began from (in relation to the sun, anyway; the sun itself is moving through space so you'd actually be millions of kilometres from your starting point, but you would still see more or less the same view of the stars). How long exactly? Well, it would be the time it takes for the Earth to complete its orbit around the sun – which is, coincidentally, just a few days more than the ever-so-convenient 360.

What better reason to divide up the rounded orbit of the Earth into 360 equal pieces? As if its ability to be factored easily was not enough, even the celestial bodies seem to be conspiring to make it the number of rotation.

The measurement of angles in a consistent way unlocked amazing secrets about our world. For instance, the ancient Greek mathematician Eratosthenes used a few simple facts about angles to calculate the circumference of the Earth with stunning accuracy.

Eratosthenes once received a letter from a friend who lived in the town of Syene, south of where Eratosthenes himself lived in Alexandria. His friend had written that at noon on the solstice, he looked down a very deep well in his town and noticed that he could see his reflection at the bottom of the well – and that it perfectly blocked out the shape of the sun behind him.

Eratosthenes was aware that the earth was spherical, despite the myth that this was a much more recent discovery. The reasoning for this was brilliantly simple: every time you observed the shadow of the Earth on the moon, it was round. The only kind of shape that casts a circular shadow from every direction is a sphere. What no one knew, though, was exactly how large a sphere the Earth was.

That mystery could be solved, however, with the help of the letter that Eratosthenes had received from his friend. He realised that his friend's observation meant that the sun was directly above Syene at noon on the solstice.

Eratosthenes put this together with another fact. At the same time, on the same day, in his own city of Alexandria, Eratosthenes observed that a pole in the ground cast a small shadow that showed the stick to be 7.2° off axis when compared to the light coming from the sun.

Perhaps this doesn't click with you as immediately as it did with Eratosthenes. It didn't with me either, so let me try and unpack the geometric thinking that he went through to reason his way through this.

You can actually follow along at home with some of this if you like – all you need are two matchsticks, some Blu Tack, a small light (like a torch, or the flash that's part of your phone), and a dark room!

Attach some of the Blu Tack to each of the matchsticks and then stand them up on a flat surface. Then close the curtains and turn on your little light. Place the light directly above your matchsticks and see if you can spot the shadow formed by each of them.

You might see some very small shadows, but as you raise your light further and further away from the matchsticks, you'll see those shadows virtually disappear. What you are simulating is noon, where the sun is directly overhead. Light rays from the sun (i.e. your torch) are hitting the matchsticks head on, causing very little shadow to form.

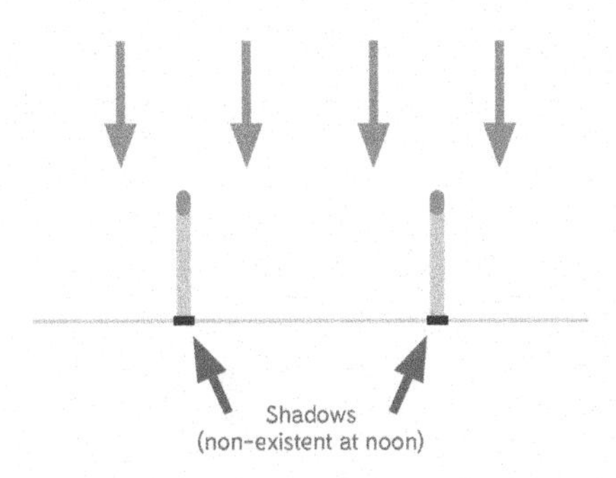

Now move the light to one side of the matchsticks, and watch as the shadows move in response. The shadows lengthen, which is something we see so frequently that it is quite intuitive – but think with me about why the shadows are getting longer. Can you see it? It's because the angle of the light rays has changed in relation to the matchsticks.

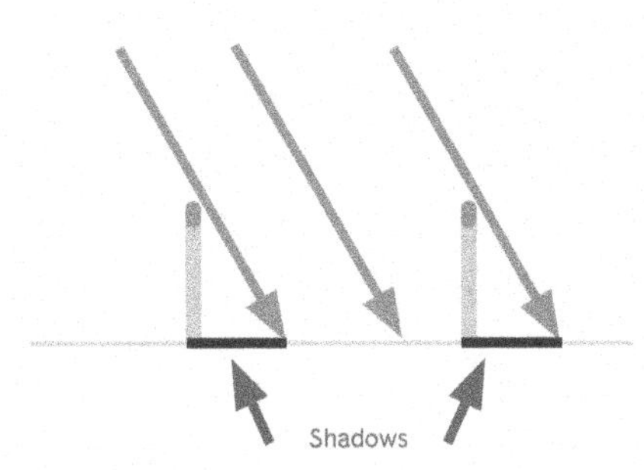

In both cases, we observe that the matchsticks create the same kinds of shadows – because the matchsticks are facing in the same direction. This will always happen if the light source is sufficiently far away, and if the surface they lie on is flat. But what if they aren't on a flat surface?

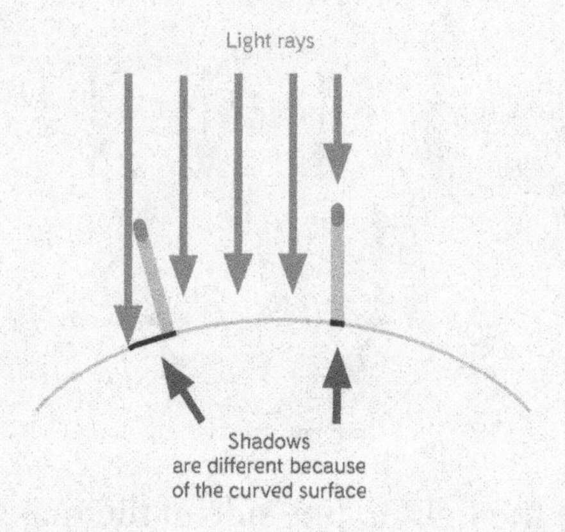

To recreate this at home I used a soccer ball, but you could use the underside of a bowl or anything curved. When the matchsticks are standing vertically on a curved surface, they are necessarily facing in different directions – and so while the same light source might be illuminating both matchsticks, one creates a shadow while the other one doesn't. This is the situation Eratosthenes realised was taking place between him and his friend: the matchstick with no shadow was like his friend in Syene (where the well was pointing directly up towards the sun), while the other matchstick was like Eratosthenes's pole in Alexandria – forming a shadow that indicated the pole was pointing slightly away from the sun.

Think about the matchsticks on the soccer ball for a minute. Both matchsticks appear vertical at their own standing points, which also means that each matchstick is pointing at the centre of the soccer ball. In the same way, since the well in Syene and the pole in Alexandria both appear vertical at their respective positions, that means they are each pointing towards the centre of the Earth.

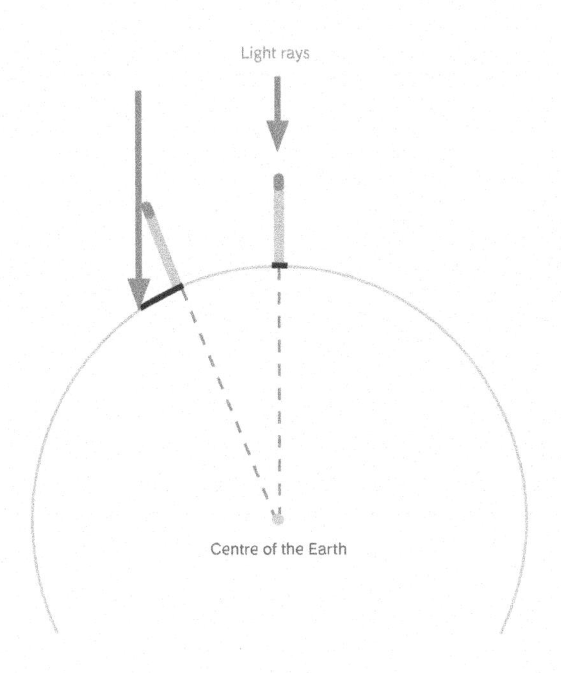

However, if we apply some geometric reasoning to this situation, we can show how this simple fact is the key to calculating the earth's circumference. The angle that Eratosthenes measured from his pole's shadow – 7.2° – is going to be exactly the same as the angle formed between the two lines that stretch to the centre of the Earth. The reason for this is that the rays from the sun (which I've marked in as AB and CD in the diagram on the next page) are parallel to each other, and 'alternate angles between parallel lines are equal' – does that ring any bells from the deductive geometry you were taught back in high school?

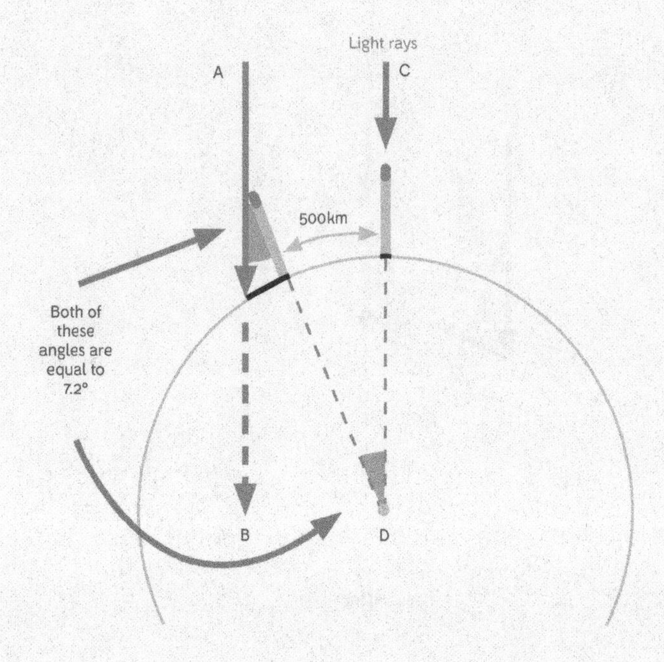

The angle of 7.2° is exactly one-fiftieth of a full revolution (360°), which means that if we know the distance between Alexandria and Syene, we just need to multiply it by 50 to get the full circumference of the earth.

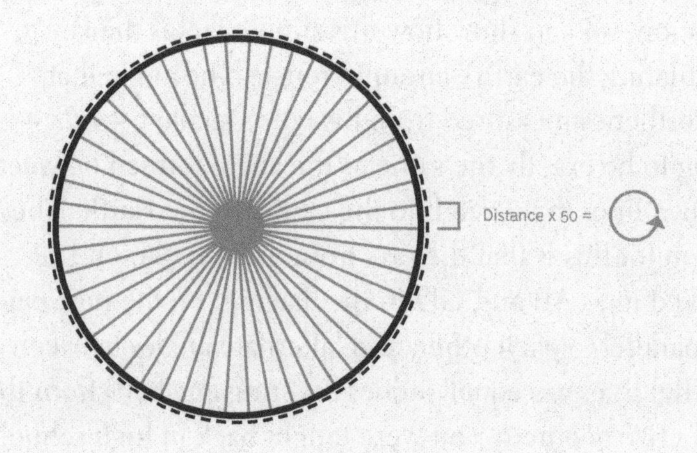

Eratosthenes did know that distance – it was a trade route that had been carefully measured by merchants of the day – and so when he took that measurement and multiplied it by 50, he came back with an answer of 44,100 kilometres. He was out by a factor of 10% – not bad considering this was more than two thousand years ago and he never even left his study to work it out.

So our closest star, the sun, helped us to quantify the size of our planet – all because of the Earth's roundness. Which quite appropriately brings us back full circle to reconsider our original question: if stars like the sun are just as round as Earth, then why do we perceive them as spiky?

The answer is as surprising as it is beautiful.

With the exception of our sun, all stars we see in the sky are so small that they appear as single pinpricks of light against the darkness. Therefore, the shape that we perceive has to do with the way the light is bent and diffracted on its journey towards us. Any objects that lie in the path of the light as it travels towards us will slightly distort the way the light ends up looking, and the distortion will depend on the shape of the object that's there. So, a vertical line will introduce a lined pattern to the light that is passing by it, as shown on the following page.

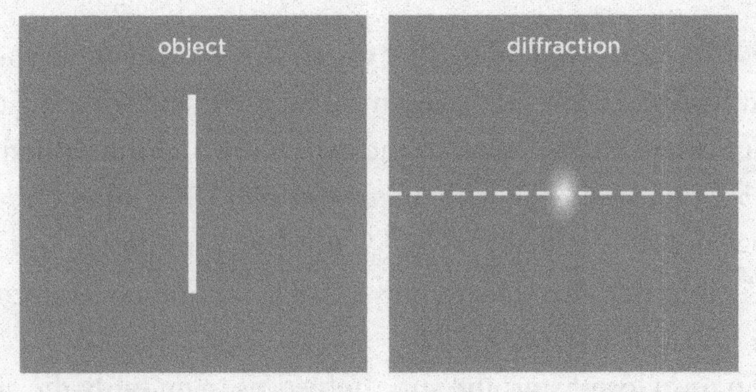

A circle will create this kind of halo effect on the light that it diffracts:

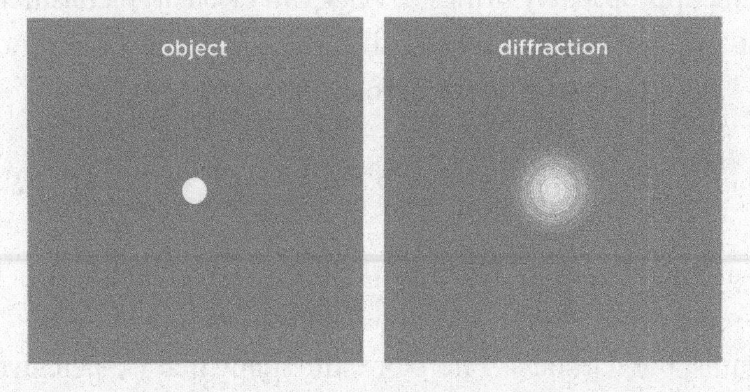

Regular polygons such as hexagons start to produce things that we might finally recognise as 'star-shaped':

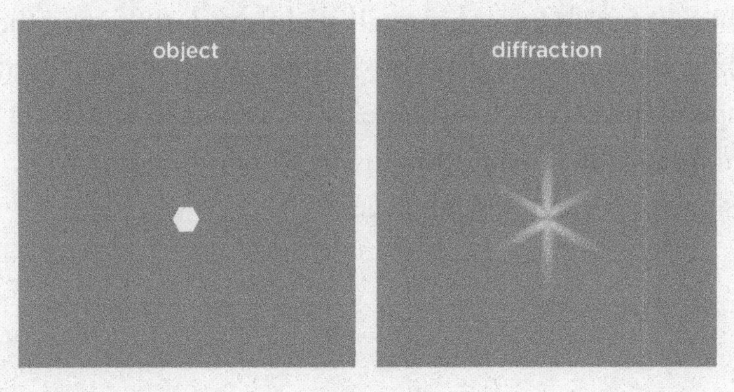

But the real kicker is: what kind of shape do things pass through if we look at the night sky and see all the stars with points on the ends? And the answer – the one thing that every bit of starlight registering in your brain comes through – is your eye. We so often forget that our eyes are not manufactured objects, but rather organically grown objects with particular structures in them. The relevant structures in this case are what ophthalmologists call 'suture lines', where the various fibres in the eye have grown together and fused. These suture lines form when the eye's muscles, growing around the spherical organs in your head, eventually meet together and join in a mess of blood vessels and other biological support structures. These suture lines are, in fact, star-shaped – and it is by seeing the world through these stars that we now see actual stars in the shape that we are so familiar with.

CHAPTER 16

DEFICIENT, PERFECT, ABUNDANT

In the previous chapter, we talked about the number 360 and how its abundance of factors made it a great choice for the number of degrees in a full revolution. The idea that some numbers have lots of factors while others have few seems like a somewhat esoteric concept, but as we saw in the 'Unbreakable locks' chapter, it turns out that there are some pretty important and useful applications of this idea to fields like cryptography. The most extreme situations of a number having few factors are the prime numbers, which can only be factorised by themselves and 1. Since the entire modern economy rests on the ability to securely encrypt and decrypt information, and essentially all of our methods for doing this rest on the use of prime numbers, it's no exaggeration to say that this field of mathematics has changed the world.

Its pragmatic usefulness to society is part of the reason why so much ink has been spilled trying to understand this area of mathematics – but, as in all areas of mathematics, practical concerns are just one piece of the pie.

An analogy might shed some light here. Human beings have mined the earth for as long as we've had tools that allow us to do it. The vast majority of such excavations have been in search of practical value: a vein of oil or a concentration of precious jewels. But some cave explorers have edged their way into the darkness with tiny flashlights simply out of curiosity. They are not just looking for useful things, they are simply wondering what they will discover: perhaps a unique geological formation, or a new animal species, or maybe even just a place of beauty. One of my favourite examples of this is the Cave of Crystals that was found in the Naica mine of Chihuahua, Mexico. It is the kind of fantastical vision that a little child might dream up lying in their bed. Except it's real.

In the same way, mathematicians probe the unknown not just looking for useful ideas (like the ability to encrypt messages, or the ability to predict the path of the planets and stars), but also out of a desire to see if something unusual or unexpected will appear. And this mathematical field – called 'number theory' – is brimming with the unusual and unexpected.

Number theory is concerned with 'the counting numbers' – that is, the infinite list of numbers that begins 1, 2, 3 and continues forever. We call them the counting numbers for an obvious reason: they're the only numbers you will ever need if all you do is count objects. There are many other kinds of numbers that arise when you start doing arithmetic: that is, when you start applying operations like subtraction, division and square roots to the counting numbers. This is how we get things like negative numbers (-1, -2, -3 . . .), or fractions (½, ¾, ⅝, . . .). But number theory steers clear of such exotic inventions, because as it turns out there is still plenty of interesting terrain to explore!

To return to the original idea of factors, it's fairly easy to demonstrate that counting numbers can be divided cleanly into three categories if we look at them through the lens of their factors.

There are the numbers that have exactly two factors, which we call the PRIMES:
2 has the factors 1 and 2;
3 has the factors 1 and 3;
5 has the factors 1 and 5;
7 has the factors 1 and 7; and so on . . .

There are the numbers that have more than two factors, which we call the COMPOSITES:
4 has the factors 1, 2 and 4;
6 has the factors 1, 2, 3 and 6;
8 has the factors 1, 2, 4 and 8;
9 has the factors 1, 3 and 9; and so on . . .

Then there is a single number which falls into neither of these categories: the number 1. It only has a single factor, namely itself. Since this category is such an exclusive club, it doesn't really have a name.

You might think it seems strange to separate 1 into its own category – and you wouldn't be alone. It's one of the most common misconceptions out there among the people who remember there are things called prime numbers; most will probably tell you that the number 1 is one of them. It's not entirely anyone's fault: the most common way to describe prime numbers is to say that they can be divided evenly by themselves and 1, which the number 1 is clearly able to do. I even stated this description myself at the start of the chapter. But a description is not the same as a definition: I can describe human beings as animals with hearts, but it wouldn't make sense to define human beings as any animals that have hearts. And the definition of prime numbers is that they must have exactly two factors – no more and no fewer.

Why do we do things this way? Why don't we just define 1 as a prime and make things look so much neater and tidier? This gets at a very profound idea that I'll address in the next chapter, 'The periodic table of numbers'.

For now, I'm interested in this divide between the primes and the composites. If you think of these as two hermetically sealed categories, then we are picturing the universe of counting numbers something like this:

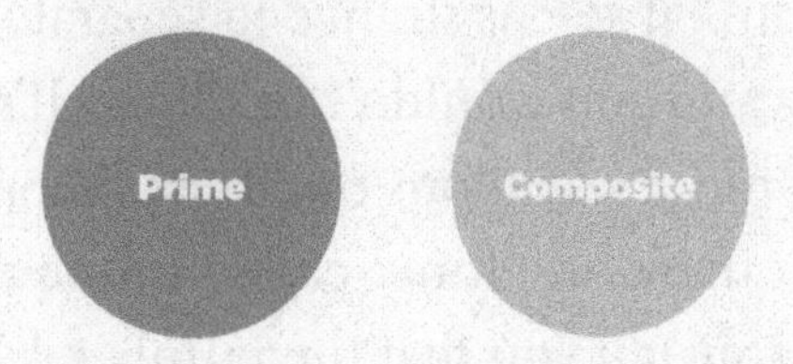

In other words, these are separate groups that have nothing to do with each other! But we started this chapter thinking about the number 360 and how it doesn't just have more than two factors – it has a lot more than two, and certainly proportionally more than the numbers around it. So this universe of numbers is starting to diversify a little – it is starting to look more like this:

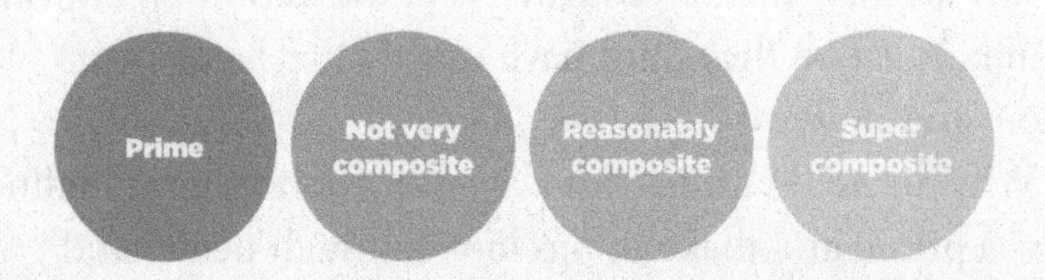

That is to say, numbers aren't just composite or not; some numbers are more composite than others and we can distinguish between those that have higher 'compositeness' (to coin a term) than others. There are lots of ways to measure compositeness, but I want to show you one of the main ways that is very simple: all you really need to know is addition and division.

Let me show it to you with the first 20 numbers. If you are keen (or want to keep a friend or relative busy for a little while), you could do this next part on your own. We want to come up with all the factors for each number – that is, what are the numbers you can divide each number by and not leave any remainders or leftovers? Here's what it looks like when you're done:

Number	Factors	Number	Factors
1	1	11	1, 11
2	1, 2	12	1, 2, 3, 4, 6, 12
3	1, 3	13	1, 13
4	1, 2, 4	14	1, 2, 7, 14
5	1, 5	15	1, 3, 5, 15
6	1, 2, 3, 6	16	1, 2, 4, 8, 16
7	1, 7	17	1, 17
8	1, 2, 4, 8	18	1, 2, 3, 6, 9, 18
9	1, 3, 9	19	1, 19
10	1, 2, 5, 10	20	1, 2, 4, 5, 10, 20

You can clearly see that some numbers appear to have a whole raft of factors where others don't. To sort out the wheat from the chaff, the next thing you do is take the factors for each number and add them up.

Number	Sum of factors	Number	Sum of factors
1	1	11	12
2	3	12	28
3	4	13	14
4	7	14	24
5	6	15	24
6	12	16	31
7	8	17	18
8	15	18	39
9	13	19	20
10	18	20	42

The more factors a number has, the larger the sum is. But larger numbers naturally have larger sums even when they don't have all that many factors, because every number is divisible by itself. Compare the number 19, which has only two factors and a sum of 20, with the number 6 which has four factors but a sum of only 12.

To overcome this, mathematicians invented this thing called the 'abundancy index' – which you can get by taking the sum of factors and dividing by the number the factors originally came from. To make things a little easier to see, we can also write each one as a percentage. Let me show you how it works with a number like 18:

The factors of 18 are: 1, 2, 3, 6, 9, 18
The sum of these factors = 39
Abundancy index of 18 = 39 ÷ 18
= 2.16666 . . .
= 216.666 . . . % (as a percentage)
= 217% (to the nearest percentage point)

Turn the page to see what happens when you work out the index for the numbers 1 to 20.

Number	Index	Number	Index
1	100%	11	109%
2	150%	12	233%
3	133%	13	108%
4	175%	14	171%
5	120%	15	160%
6	200%	16	194%
7	114%	17	106%
8	188%	18	217%
9	144%	19	105%
10	180%	20	210%

From this we can rank the first 20 numbers in terms of their 'abundancy index' – here's the list from highest to lowest:

12, 18, 20, 6, 16, 8, 10, 4, 14,
15, 2, 9, 3, 5, 7, 11, 13, 17, 19, 1

There are some really important things to observe here. First, notice how all the numbers at the front of the list –

all the numbers with a high abundancy index – are even. In fact, with the exception of the number 2, the list neatly divides into two halves – the even numbers on the top row and the odd numbers on the bottom. But in addition to that, there is a really important list of numbers over there on that bottom row – in fact, you can see all the prime numbers, in perfect ascending order, in the second half of the list (except for the number 1).

So far we have been using this thing called the 'abundancy index' to measure our numbers. This is because any numbers with an index above 200% are called 'abundant' numbers, while numbers with an index less than 200% are called 'deficient'. Under this scheme, only three of the first 20 numbers are abundant: 12, 18 and 20. A single number, being 6, is called 'perfect' because its index is neither above nor below 200% – it is exactly equal. This, then, is what the 'continuum' of numbers looks like when they are considered in terms of their factors:

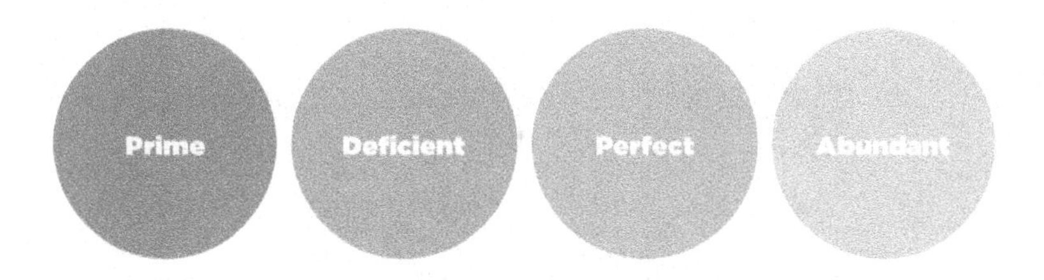

Now that we've established a way to quantify whether a given number has lots of factors or not very many, we can return to the number that started us off on this journey in the first place – 360. It must clearly be an abundant number, but just how abundant is it? Well, let's crunch the numbers. First, we find the factors of 360 and add them up:

$$1 + 2 + 3 + 4 + 5 + 6 + 8 + 9 + 10 + 12 + 15 + 18 + 20 + 24 + 30 + 36 + 40 + 45 + 60 + 72 + 90 + 120 + 180 + 360 = 1170$$

Then we divide this sum (1170) by the original number (360) and see what we get:

$$1170/360 = 325\%$$

Wow – this is way higher than anything we saw in the first 20 numbers. In fact, as it turns out, this is higher than anything in the first 1000 numbers – it is the most abundant number, no exceptions, between 1 and 1000. If there was such a thing, 360 would be in the category of 'super-abundant' numbers!

CHAPTER 17

THE PERIODIC TABLE OF NUMBERS

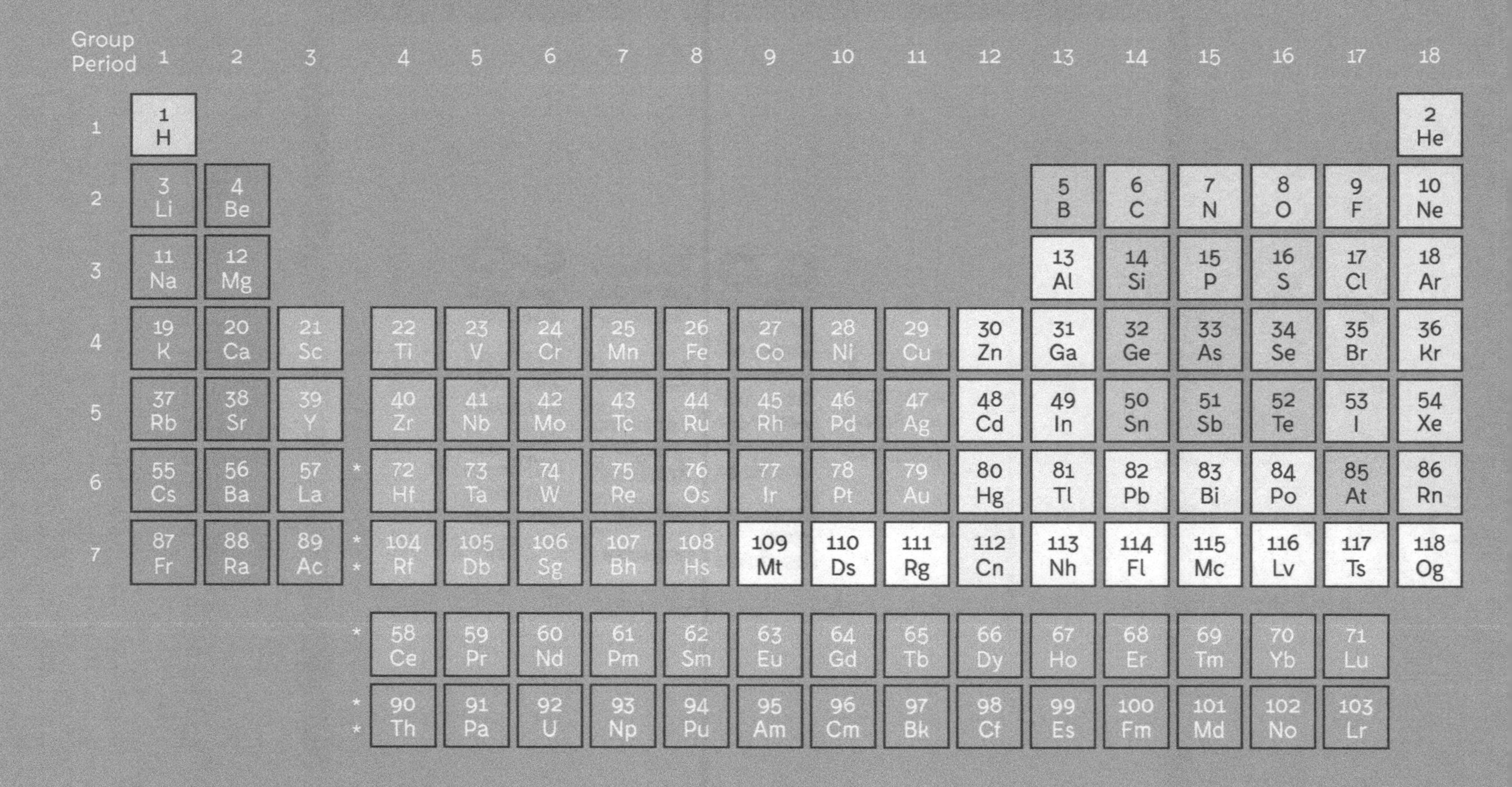
Group
Period
1 2 3 4 5 6 7 8 9 10 11 12 13 14 15 16 17 18
1
1 H
2 He
2
3 Li
4 Be
5 B
6 C
7 N
8 O
9 F
10 Ne
3
11 Na
12 Mg
13 Al
14 Si
15 P
16 S
17 Cl
18 Ar
4
19 K
20 Ca
21 Sc
22 Ti
23 V
24 Cr
25 Mn
26 Fe
27 Co
28 Ni
29 Cu
30 Zn
31 Ga
32 Ge
33 As
34 Se
35 Br
36 Kr
5
37 Rb
38 Sr
39 Y
40 Zr
41 Nb
42 Mo
43 Tc
44 Ru
45 Rh
46 Pd
47 Ag
48 Cd
49 In
50 Sn
51 Sb
52 Te
53 I
54 Xe
6
55 Cs
56 Ba
57 La
*
72 Hf
73 Ta
74 W
75 Re
76 Os
77 Ir
78 Pt
79 Au
80 Hg
81 Tl
82 Pb
83 Bi
84 Po
85 At
86 Rn
7
87 Fr
88 Ra
89 Ac
* *
104 Rf
105 Db
106 Sg
107 Bh
108 Hs
109 Mt
110 Ds
111 Rg
112 Cn
113 Nh
114 Fl
115 Mc
116 Lv
117 Ts
118 Og
*
58 Ce
59 Pr
60 Nd
61 Pm
62 Sm
63 Eu
64 Gd
65 Tb
66 Dy
67 Ho
68 Er
69 Tm
70 Yb
71 Lu
* *
90 Th
91 Pa
92 U
93 Np
94 Pu
95 Am
96 Cm
97 Bk
98 Cf
99 Es
100 Fm
101 Md
102 No
103 Lr

The nineteenth century was an exciting but confusing time to be a chemist. New scientific inventions like the battery and the spectroscope enabled a string of new elements to be discovered in relatively quick succession. There was a clear sense of progress as the library of humanity's chemical knowledge grew in size, but there was also a sense of unease as the library's shelves seemed to be growing haphazardly and without order. Science was supposed to be characterised by order and simple laws, but chemistry was becoming an unruly mess.

Enter Dmitri Mendeleev, born in 1834 in the Siberian village of Verkhnie Aremzyani. More than 2000 kilometres from Moscow, it must have seemed an unlikely place to grow a scientific genius. But with a single thunderclap of insight, Mendeleev transformed the way that people thought of the substances that make up the universe all around them. Chemists in Mendeleev's time knew of roughly 60 elements – about half of the ones we know about now. However, there was little consensus as to why each element had the properties that were observed. Why were some elements able to conduct electricity effectively, while others couldn't? Why did some have high boiling points while others didn't? No one really had a particularly satisfying answer that made sense of all the data.

That is, until Mendeleev. His flash of inspiration was that when the elements were ordered by their atomic mass, the properties of the elements seemed to come in cycles. Mendeleev didn't have the complete picture back then, but with our modern understanding of atomic nuclei – and their positive subatomic particles, the protons – we can now understand the cyclical properties that Mendeleev observed.

Lithium (three protons), sodium (11 protons) and potassium (19 protons) are all highly reactive substances – so much so that they can even explode when exposed to water. By contrast, helium (two protons), neon (10 protons) and argon (18 protons) are so unreactive that they are named the 'noble gases' since they interact so little with the other elements! In each case, we see that similar properties are shared by elements that are separated by eight protons. The pattern does become a little more complicated later on, but the essential idea remains the same: chemical properties are periodic – they repeat predictably as we consider atoms with more and more protons than their lighter cousins.

Mendeleev arranged the elements that he knew about into a table that grouped these similar elements into columns. What he recognised was that there were gaps in his table, indicating that there were elements which humanity had not yet discovered but that his model predicted should exist – and he could even anticipate what chemical properties they should have.

This is how the periodic table that we know today was born.

Chemistry benefits from the insights of mathematics. From the numbers of protons, neutrons and electrons in each element to the periodicity of behaviours and properties that Mendeleev codified in his famous table, there are many aspects of chemistry that are best understood by looking through the lens of a mathematician. But something that is often overlooked is that the reverse is also true. Mathematics also benefits from the insights of chemistry – especially when you look at the parallel between elements and prime numbers.

Many of the substances that we interact with on a daily basis are chemical compounds, not elements. Elements are things like carbon, oxygen and hydrogen. Compounds are things like water, methane and ethanol – which are made by combining those elements in different quantities and arrangements. Water is famously made of two hydrogen atoms and one oxygen atom. Methane has one carbon atom and four hydrogen atoms. Ethanol, an alcohol, has two carbon atoms, one oxygen atom and six hydrogen atoms. Other compounds have more obvious titles that literally state their composition: the famous greenhouse gas carbon dioxide has, as its name suggests, a single carbon atom and two oxygen atoms.

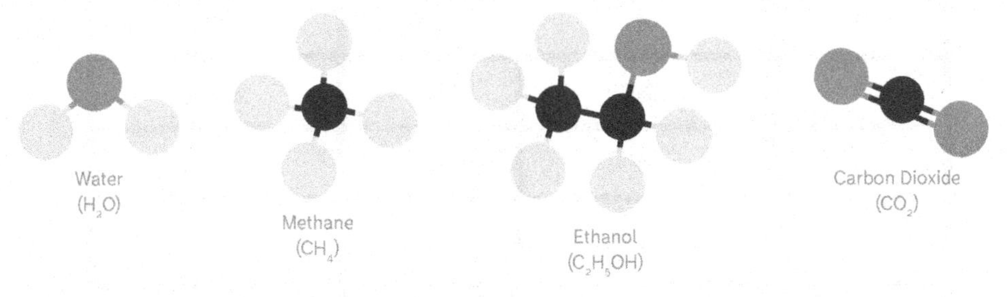

In the same way that you can combine elements to form compounds, you can combine prime numbers to form what we call composite numbers.

In the chapter 'Deficient, perfect, abundant', we talked about composite numbers as numbers that have more than two factors, but this is actually a bit of a roundabout way to understand them. A more insightful way to think of them is like numerical compounds: they are what happens when you combine the prime numbers together.

Let me show you what I mean. We need to return to the idea of 'factorisation' that we mentioned in the previous chapter. If we think of the first few composite numbers, we can factorise every single one using just prime numbers. This is called a number's 'prime factorisation'. As before, let's look at the numbers 1 to 20:

Number	Prime factorisation	Number	Prime factorisation
1	1	11	11
2	2	12	$2^2 \times 3$
3	3	13	13
4	2^2	14	2×7
5	5	15	3×5
6	2×3	16	2^4
7	7	17	17
8	2^3	18	2×3^2
9	3^2	19	19
10	2×5	20	$2^2 \times 5$

Just like every chemical compound has a unique structural formula that shows the various elements it's made up of, every composite number has a unique prime factorisation that shows the prime numbers it's made up of.

This idea is so important that it gets a very fancy name: it's called the Fundamental Theorem of Arithmetic, and here it is in all its flowery formality:

> Every integer greater than 1 must be either a prime number itself, or able to be uniquely represented as the product of prime numbers.

All this means is that starting from 2 and going upwards forever, any whole number is either prime, or can be written by multiplying a combination of prime numbers together. And this combination is always unique.

The 'uniqueness' of this prime factorisation is an important and in some ways unexpected quality. It's unexpected because, as we saw in the last chapter, composite numbers can frequently be factorised in many different ways. For instance, 84 can just as easily be written as 4×21, 6×14 or 7×12. However, as you can see on the next page, if we continue factorising until we are left with nothing but primes – we always, always end up with exactly the same primes. Every number has a unique prime factorisation, just like every compound has a unique molecular formula that describes it.

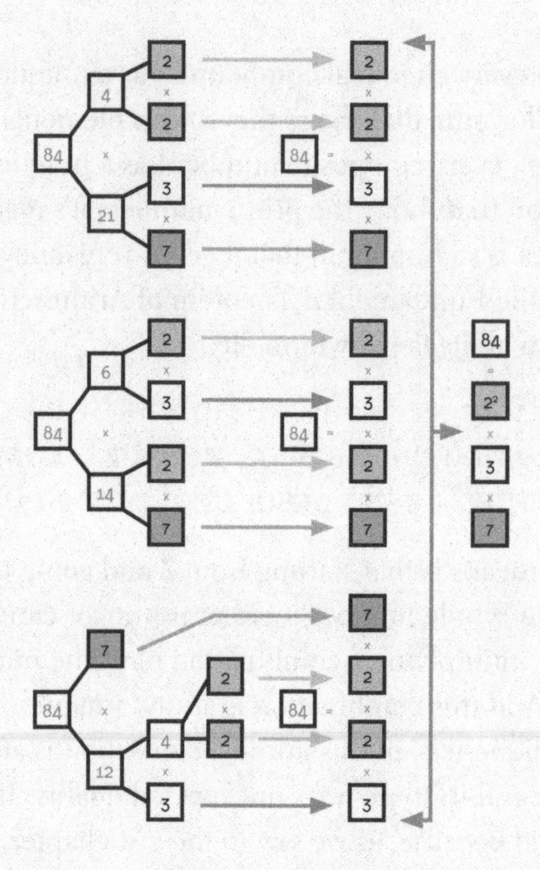

This, funnily enough, is actually the main reason why we don't define 1 as a prime number. If 1 were a prime number, the Fundamental Theorem of Arithmetic would break down – because there would not be a unique prime factorisation of 84. In addition to $2^2 \times 3 \times 7$, I could (rightly) say that $84 = 1 \times 2^2 \times 3 \times 7$, just as I could say that $84 = 1 \times 1 \times 2^2 \times 3 \times 7$, and so on. There would be an infinite number of prime factorisations for each number, rather than just one for each.

You can also state this in reverse. When it comes to the periodic table of elements, it's easy to know when you're talking about one element as compared to another. The thing that fundamentally matters is the number of protons. Every atom in the universe with six protons is carbon. You could take one of those carbon atoms and add or remove electrons: you'd end up with an ion, which is slightly different but still carbon. You could take another and add or remove neutrons: you'd end up with an isotope, which is slightly different again but still carbon.

Add a proton or two, though, and you've got a whole new ball game. Add one proton and you'd get nitrogen, which behaves wildly differently from carbon. Add two protons and you've got oxygen. That process is so difficult that it only really occurs in the heart of a star where there are literally astronomical amounts of heat and pressure – the necessary conditions for nuclear fusion to occur.

The parallel to this idea is that when you multiply by a prime number, you get a whole new composite number. Just like changing carbon into oxygen, multiplying 3 by 7 gives you 21 – a whole new number with all new properties. But if 1 were prime, wouldn't that be strange? You can multiply 3 by 1 as many times as you like and you would still end up with the same number you started with: 3. It's hardly worthy of the name 'prime' if it makes no fundamental change to numbers no matter how many times you multiply by it!

There's an infinite universe of numbers out there – and thankfully you don't need to be in the heart of a star to make up new ones. All you need is the furnace inside your mind to forge as many numbers as you can imagine.

CHAPTER 18

CONSPIRACY THEORY

In the early 1970s, the Watergate scandal engulfed US politics and radically transformed the way people viewed the presidency. Richard Nixon, the sitting US president, was found to be guilty of seriously breaking the law and staging a massive cover-up of the entire affair. Part of why this event had such a deep and lasting effect on the popular psyche was because it went through such a change in the eyes of the public.

When news about Watergate first broke, most people passed it off as an unimportant incident that would disappear out of the news cycle just like every other piece of trivia that was reported in a normal day. As the investigation deepened and people realised that this incident wasn't going away in a hurry, the popular perception was that Watergate was becoming an unfounded witch hunt. Few people could stomach the idea that the leader of the free world was a criminal who had wilfully participated in treasonous acts and abused his executive powers to obstruct justice. For a time, the few people who believed that President Nixon was guilty were viewed as cranks, conspiracy theorists who had concocted a nonsensical fiction that couldn't possibly be feasible, let alone true.

That was, until the entire plot unravelled and suddenly everyone's worst fears were confirmed. Through an astonishing series of revelations that would rival the twists and turns in a Hollywood blockbuster, the truth was revealed – and the conspiracy theorists had been right all along.

Watergate marked a watershed in the era of conspiracy theory. Until then, those who believed in secret codes, clandestine societies and government cover-ups were routinely regarded as crazy. But Watergate forced everyone to admit that even the most implausible theories could sometimes be true.

Mathematics tells its own story about conspiracy theories and why they continue to crop up in every corner of the world. Surprisingly, it helps us to see that there will always be grist for the conspiracy theorist's mill. The nature of our world and the self-made ocean of data we swim in ensure that there will always be material for theorists to point to as 'evidence' of suspicious behaviour and secrets being dangled in front of our noses.

To understand what's going on here, it will help to think of a very simple children's puzzle that you have almost certainly engaged with at some point in your life:

the find-a-word.

I used to spend hours playing find-a-word puzzles. Growing up, I actually had a book full of them (which I now realise was one of my mum's key strategies for getting some time to herself while I was distracted).

Here's an example of what one of the puzzles might have looked like:

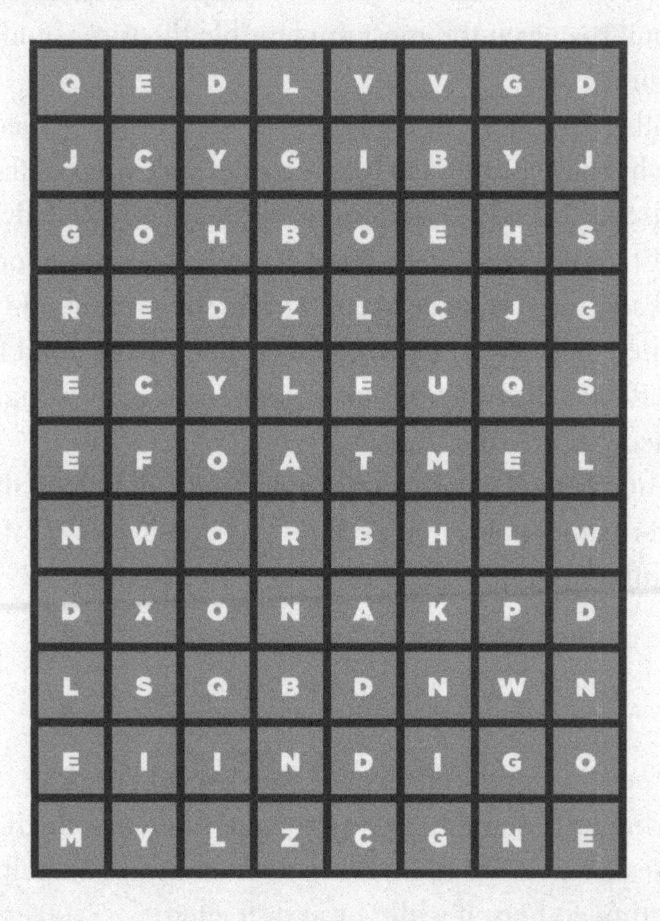

This find-a-word contains all seven canonical colours of the rainbow: red, orange, yellow, blue, green, indigo and violet. Words may be hidden vertically, horizontally or on diagonals. Watch out – some are written in reverse (right to left, rather than left to right). For a bonus, there is one other colour hidden inside the grid that I haven't named – see if you can spot it!

It isn't hard to create your own find-a-word. The way that I created this one was simply by starting with an empty grid and gradually adding the words that I wanted to hide. Once I was done, it was just a matter of filling in all the gaps with a series of random letters. Hey presto! It's a find-a-word.

What would happen, though, if I created a find-a-word and skipped that first step – the one where I added in my own words? What if I constructed a find-a-word that was filled completely with random letters? Here's an example of what happens if I do exactly that on a 3 × 3 grid:

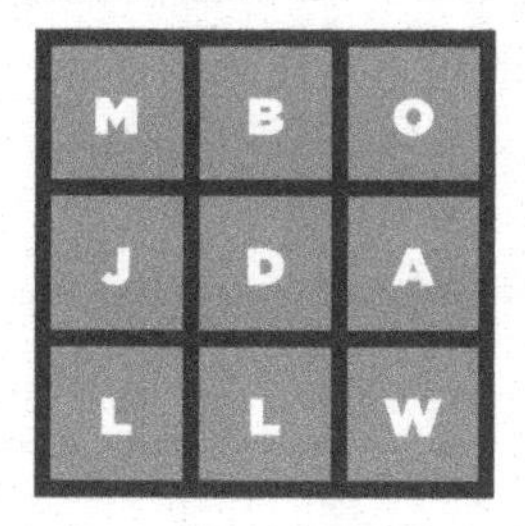

Unsurprisingly, it looks pretty much like gibberish. That's exactly what it is – there are no English words to be found in this grid, no matter which way you look. But now see what happens if I add an extra row and an extra column:

If you start from the letter B on the first row and follow diagonally downwards and to the right – reading clear as day, you can see that the word 'bad' has appeared! Not only that – on the second row, starting at the second letter from the left and reading left to right, there's another word: 'dad'. Is this random collection of letters trying to tell me that I'm not a very good father?!

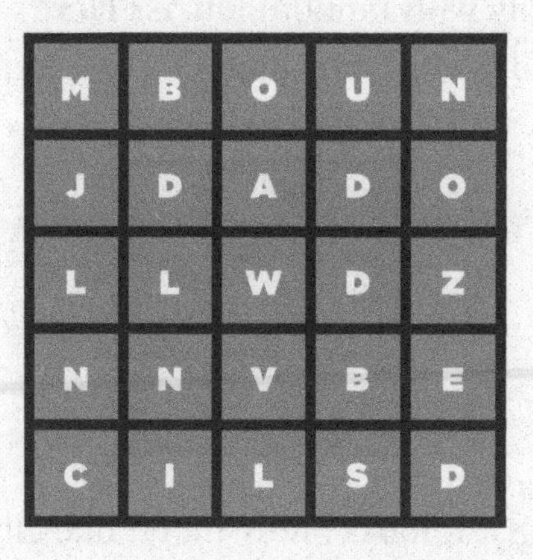

If I enlarge the grid one more time to create a 5 × 5 grid, you see even more words spontaneously appearing. In addition to 'bad' and 'dad', I can spot 'in', 'do', 'no', 'zed', 'be', 'bade' and 'ado'. These words are multiplying!

What on earth is happening here? I'm not making this up or surreptitiously smuggling words into these grids manually – here are two more grids that were randomly generated in the same way, and each is similarly filled with words:

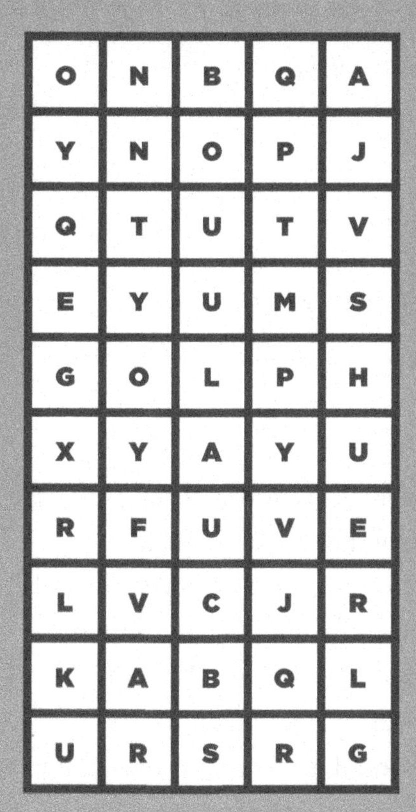

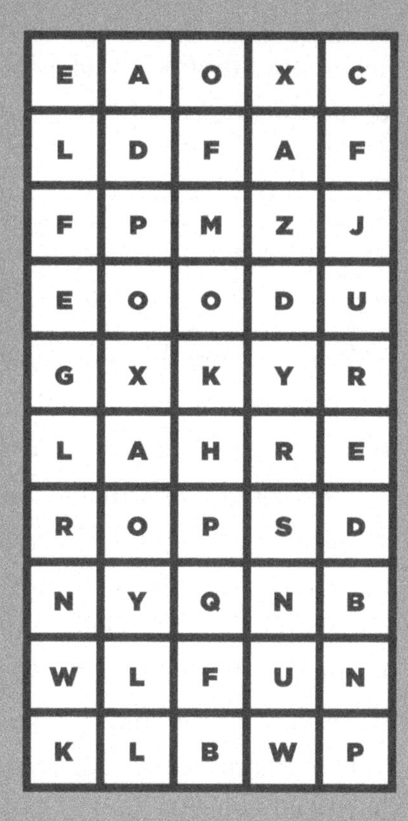

On the left grid, I can see: pony, ox, log, not, hue, yay and spa.

On the right grid, I can see: elf, ox (three times!), ex, cam, do, red, fun, gap, oh, lo and no.

You can create your own 'random' find-a-word grids by typing the following web address into your search engine: **www.bit.ly/findaword.**

If you use it and have a play, you too will find that it is actually remarkably difficult to make a grid that is completely devoid of English words. So what is going on here?

The phenomenon that you're experiencing is the impossibility of disorder.

In a sea of chaos, there will always be islands of order – as long as the sea is large enough.

We illustrated this with the 3 × 3 grid that had no words – but as soon as it starts to get larger, it becomes very difficult to avoid the creation of words.

The area of mathematics that deals with this situation is called Ramsey theory, named after the British mathematician and economist Frank P. Ramsey. Very few of you are likely to have encountered this kind of mathematics because it is in a field called graph theory, which is not part of any compulsory mathematics course in Australia.

If the mathematics taught in school is like a tour of Sydney, then algebra is like the Opera House – everyone sees it eventually. Graph theory, on the other hand, is like your local convenience store – compared to a tourist destination, only a small group of people know about it. And just like the convenience store, those people know it because it helps them solve the problems they encounter on a daily basis.

The clearest demonstration of Ramsey theory is a scenario known as the Party Problem. Anyone who's ever thrown a party knows the challenges of working out who to invite. Of course all the people you invite will be friends with you, but what about the friendships they may or may not have with each other? Graph theory is the mathematics of connections, and it helps us to understand any situation where things are related to each other through specific links. This might be suburbs linked by rail lines, or houses linked to electricity lines – or people linked by friendships.

Suppose, for instance, you want your party to have at least three people who will all be either mutual friends or mutual strangers. In either case, you can guarantee that they can have an interesting conversation. If they all know each other, they can get on like a house on fire. If they're all meeting for the first time, then they can all get to know each other from scratch at your party. Happy times guaranteed!

The first way that graph theory helps us understand and unpack this problem is that it gives us a way to represent what is going on. Some questions in life are hard to answer because it is challenging to even understand what is going on. But if we can show the essential details of a situation in a simple diagram, we are halfway to working out a solution.

The two circles below represent two people at the party. If there is a solid line between them, they know each other; if there is a dotted line, they don't know each other.

A and B are friends

A and B are strangers

Once we can use these tools, we can draw diagrams to represent different types of 'parties' where various people know or do not know each other. If we invite five other people to the party, then there are 12 basic combinations of relationships between one guest (in this case A) and the other guests, as shown below. In each of the situations, you can see there is a group of three highlighted in red – these are either mutual friends or mutual strangers.

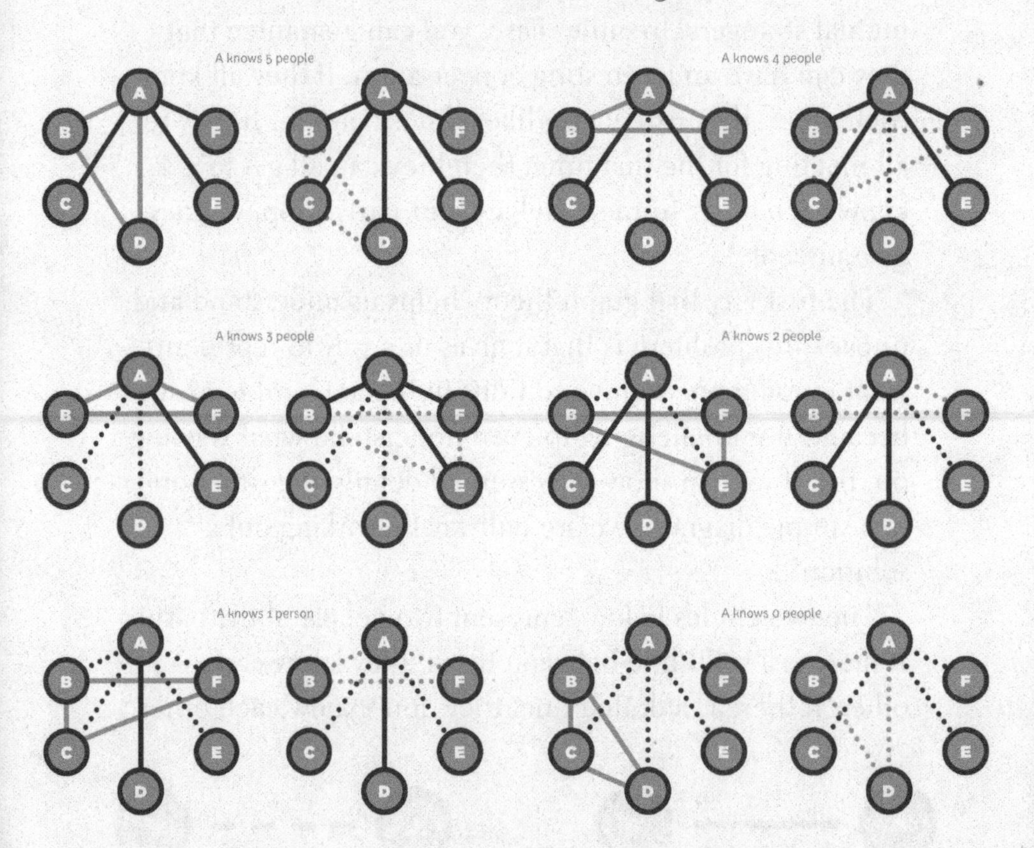

Since we can always find a group of three who either all know each other or all don't know each other, this means that if you invite at least six people you will always find such a group. If you are interested in how we can form the diagrams on the left and prove that six is the smallest number where this happens, skip forward to the chapter titled 'Six's a crowd'.

What relevance does this have to our find-a-word – or our conspiracy theorists, for that matter? Well, Ramsey theory tells us that as you increase the size of a 'combinatorial structure' – which might be a group of friends, or a find-a-word grid, or even a newspaper article – then certain kinds of structures and 'patterns' are guaranteed to appear. This is why English words seemed to pop up out of nowhere once our grid grew beyond a certain size.

This is, in turn, how many conspiracy theories begin. Keen eyes that are searching for a pattern will always be able to find something that looks suspicious if their data set is big enough.

One group of people who routinely search for numerical patterns are numerologists, people who specialise in ascribing cosmic meaning and significance to the occurrence of particular numbers. Numerologists had a field day in 2017 when the artist Jay-Z released the album *4:44*, whose title track is named after the time in the morning when the singer-songwriter apparently woke up and composed it.

One numerologist's response to the song referenced deep connections to Jay-Z's personal life: '[His wife's] birthday is on the 4th, his mother's birthday is on the 4th, his own birthday is on the 4th and he got married on the 4th.' That is certainly a striking pattern, but Ramsey theory assures us that this kind of freak occurrence is guaranteed to happen completely by chance somewhere in our world of seven billion people. (After all, there are 12 days in the year when it is the 4th day of the month – that means we ought to expect that more than 230 million people currently alive were born on the 4th, and chances are that quite a few of those people married each other!)

Ramsey theory, and its assertion that structure must appear spontaneously out of chaos (as long as there is enough chaos to work with!), pops up in some of the most unexpected ways in everyday life. For instance, Apple ran into a surprising instance of spontaneous structure when it released its iPod into the world. Though there had been portable music players for years before, the iPod marked a significant increase in the number of people who suddenly had access to their entire music library wherever they were. No longer limited to having a single album loaded into their CD player, people could easily access hundreds or even thousands of songs in their pockets.

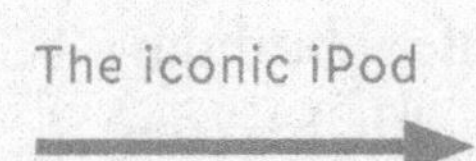
The iconic iPod

With this newfound ability, the 'shuffle' feature of the iPod came under sudden scrutiny. It was designed to randomly select songs from the music loaded onto the device, and it did as it was told – but users around the world started to report strange behaviour. 'My iPod is playing up – I have music from dozens of artists on there, but every now and then it will just play four or five songs by the one band all in a row!' People felt as though their devices were malfunctioning, and not 'shuffling' the music as advertised. Others theorised that their specific device seemed to have its own personality, complete with preferences for some artists over others. 'Why don't I ever seem to hear my Madonna songs . . . my iPod seems to be obsessed with The Foo Fighters!'

With our knowledge of Ramsey theory up our sleeves, we recognise that this is in fact exactly what we should expect. If we listen to randomly shuffling songs for hundreds of hours on end, then at some stage we ought to expect to get a streak of several songs from the same artist. The longer we listen, the more likely the streak becomes – just like making the find-a-word grids larger made it more likely that an actual word would be spontaneously generated.

Ironically, the reason we think we see patterns in randomness is largely because we humans actually have a very poor sense of what it really means for something to be random. For instance, consider the following sequence of heads and tails generated by flipping coins:

1	H	11	T	21	H
2	H	12	T	22	H
3	H	13	H	23	T
4	T	14	H	24	H
5	H	15	H	25	T
6	T	16	H	26	T
7	H	17	T	27	T
8	H	18	T	28	T
9	T	19	H	29	H
10	H	20	T	30	H

Now compare it to this list:

1	T	11	H	21	H
2	H	12	T	22	H
3	H	13	H	23	H
4	T	14	T	24	T
5	H	15	T	25	H
6	T	16	H	26	T
7	T	17	T	27	T
8	T	18	H	28	H
9	H	19	H	29	T
10	H	20	T	30	H

Spoiler alert: one of these lists was not actually created by flipping coins – it was made up by a human being who was pretending to flip coins. Can you tell which is which?

Mathematics can tell – because coin flips are so simple to understand in terms of probability and chance, we can accurately predict how often different sequences of flips (such as a head followed by a tail, or three tails in a row) should appear. The truth is that the first list came from real coin flips, while the second list was faked. The second list gives away its human origin because of how seldom it contains long consecutive streaks. Humans don't think that flipping four tails in a row or four heads in a row is normal – but if you flip a coin enough times (as demonstrated by the first list), it is virtually inevitable that this will happen at some stage.

The British mentalist and illusionist Derren Brown took advantage of this phenomenon to great effect by flipping ten heads in a row on live camera for one of his shows. In this age of computer-generated special effects, most people assume that this must be a television trick. But there are no cuts or tricks involving deceptive camera angles – he really does flip ten consecutive heads in one continuous shot. To achieve that shot, though, it took them over nine hours of filming flip after flip in order to get the full sequence! This might sound extreme, but persevering for so long virtually guaranteed the occurrence of a successful streak. For instance, turn the page and see a list of 2025 flips which contains a streak of not ten but 15 heads in a row!

There can be more serious implications to this mathematical reality. When it comes to matters of chance, human instinct is that long continuous streaks of any one event are quite unlikely. When this instinct goes wrong in a casino, the effects can be disastrous. People who become addicted to gambling often report that they could not shake the feeling that after so many losses in a row, they were bound to get a win eventually. This is called the gambler's fallacy because it is as untrue as it is tragic – those convinced of the fallacy almost universally find themselves out of cash because their instincts turn out to be so wrong.

Things become even more nuanced when this aspect of Ramsey theory interacts with human physiology. The well-known placebo effect refers to when people's health is improved by taking a medicine or undergoing treatment that has no relevant biological effect. It is so well documented that clinical trials of new medications are required to have a placebo group alongside the control group and experimental group. While the control group takes no medication and the experimental group takes the medication being tested, the placebo group takes sugar pills that contain no active ingredient – but they are told that they are taking genuine medication.

Without exception, some members of the placebo group will report that they feel better because of the medication – the one that they aren't actually taking. For some of them, their improvement is linked to the fact that they believe that they're receiving actual medicine. How can the human body recover and improve because of something that isn't even real?

The power of patterns is at work. Human beings in the modern age have been trained from childhood to associate medicine with health. In psychology, this is called classical conditioning. Conditioning leads to real biological effects, most memorably demonstrated by the Russian physiologist Ivan Pavlov. In his famous experiment, he would feed a group of dogs only after ringing a bell. After establishing this pattern, the mere act of ringing a bell – even in the absence of food – would make the dogs salivate in anticipation of their meal. Statistically, we could say that Pavlov was giving the dogs data to suggest a kind of correlation between food and the sound of the bell.

Let's put this idea together with Ramsey theory. Imagine that you started selling sugar tablets with no medicinal effect, but labelled them as cold-and-flu medication. People start buying them when they are sick, since it is a new and interesting product that they feel is worth trying. From what we know of Ramsey theory, we can predict that if there is a sufficiently large group that buys and takes the tablets, then there will be a group of people who randomly experience a reduction in their symptoms or a quicker recovery from illness at the same time as they take the tablets. These consumers may then unwittingly condition themselves to actually become somewhat better when taking placebo tablets that have no real effect!

Once you know how to look, these islands of order amid the chaos start turning up everywhere. The sky is a perfect example of a large data set that is bound to produce all kinds of unusual patterns. This is true in the day, when fish sometimes swim through the air instead of the sea:

But it is more prominently true at night, when the stars provide more than enough opportunities for people to find interesting shapes and stories among them, as they have through the ages.

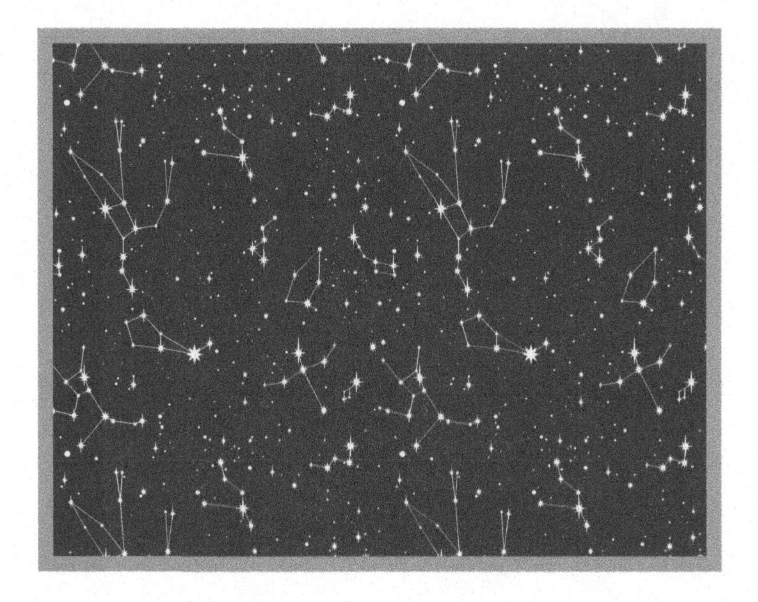

CHAPTER 19

WHAT IS PROOF, ANYWAY?

In the previous chapter, we mentioned the idea of graph theory and briefly introduced the Party Problem. I mentioned that if you want to have at least three mutual friends or three mutual strangers at a party, then you must invite at least six people. But how do we know this is true? How can we prove that five is not enough and seven is more than is necessary?

This is a wonderful opportunity to explore a really big concept in mathematics: the idea of proof.

To 'prove' something means to provide an argument or evidence for whether something is true or not.

Mathematicians aren't the only people who prove things, of course – but it turns out that 'proof' means very different things depending on your context.

For instance, consider the notion of 'scientific proof'. This has been the keystone of human progress since the Enlightenment, and without it we would arguably still be living in the Dark Ages. The scientific method is founded upon the idea of experimentation and repeatable observations. If a phenomenon can be demonstrated reliably under given conditions and reproduced by others, then most people would agree that this would constitute scientific proof of the hypothesis being tested.

However, there are countless things in life that can't be subjected to this process. History is an easy example to point to: by its very nature, it cannot be repeated. So how can we 'prove' that anything in the past ever happened?

Since experimentation is out of the question, historians and archaeologists have established clear hierarchies of evidence that can be used to verify whether some historical idea is true or not. We consider things such as printed materials, eyewitness testimonies, independent sources and physical artefacts when trying to work out the most accurate theory that accounts for all the evidence in a convincing way.

Both the scientific method and the historical method, however, have some critical weaknesses. These can't be avoided – they are essential to the nature of scientific and historical knowledge – but they are real nonetheless.

We can summarise the problems for both fields in one phrase: the lack of comprehensive knowledge. When it comes to science, we are often restricted in what we can know due to the limitations of our instruments. It is as though we are trying to look at our world through a small keyhole and we can't see everything that's happening on the other side, so we have an incomplete picture. As new technology is devised that can perceive our world with greater detail, experiments can be redesigned to take advantage of these tools and new realities are often discovered that contradict earlier beliefs. Better technology widens the keyhole – and in some cases it blows open the whole door.

This is not a bad thing – it is how we make scientific progress! A wonderful illustration of this is the way that our understanding of the atom has advanced over time. The word 'atom' literally means 'indivisible', and scientists originally proposed it as a name because atoms were so fantastically small that no one could ever have imagined them being subdivided into smaller objects. But when the physicist J. J. Thomson devised a way to measure the mass of what we now know as electrons, we were forced to revise our previous models in light of new knowledge. Thomson proposed that electrons were situated in the atom like this:

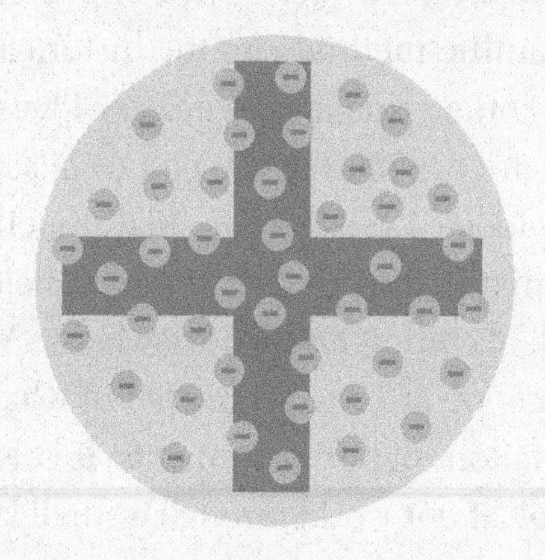

He pictured negatively charged electrons evenly distributed throughout the atom like fruit in a dessert dish, which is how his idea became affectionately known as the 'plum pudding model'. But things changed again when Ernest Rutherford directed an experiment in which he showed that in fact there was a densely packed core of matter in the middle of the atom, around which the electrons orbited. Rutherford named this core the 'nucleus', and in so doing achieved two things: he became the father of nuclear physics, and also birthed an image so striking that it would come to symbolise the entirety of science itself:

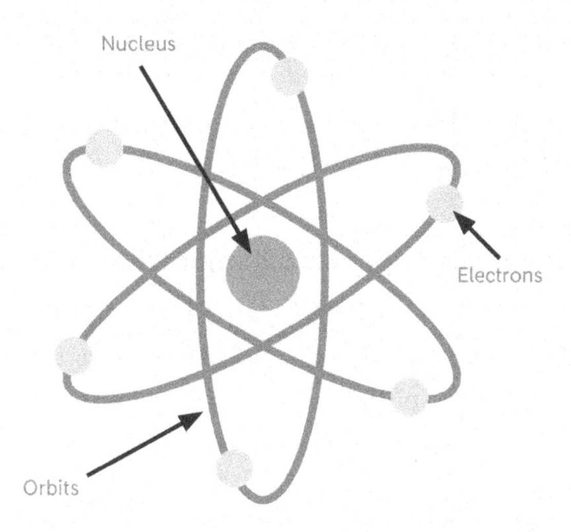

Better instruments would change this model again, though. Through the use of more high-powered equipment, we have gained even further insight into the atom, and physicists now speak of electrons inhabiting a 'cloud' where the atoms fly around in interestingly shaped zones called orbitals, depending on how much energy they possess:

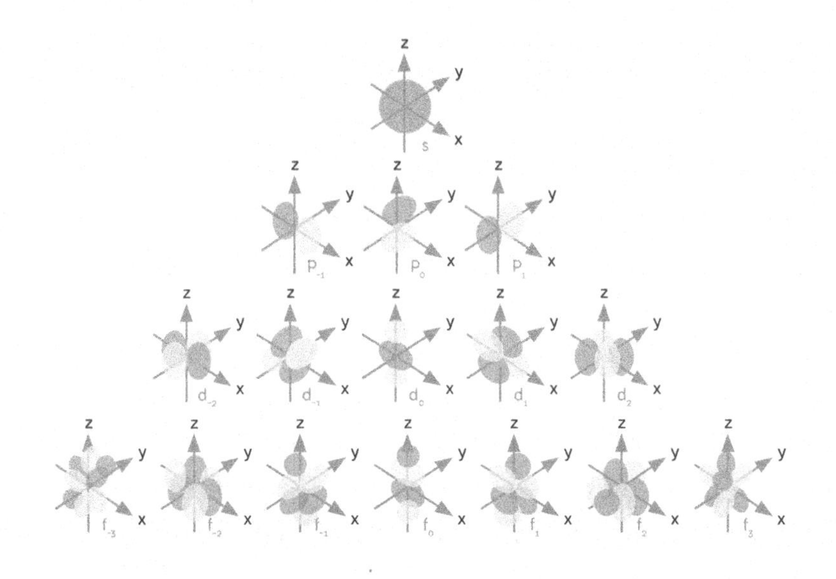

As science has progressed, we have constantly had to revise models that had once been 'proven' – now that we have a clearer idea of what is happening through better technology.

Historians face a similar dilemma, though for a slightly different reason. Our knowledge of the past is also incomplete because it is often buried in the ground, or concealed because the accounts of eyewitnesses either never existed or have been since destroyed. Occasionally, though, we make new discoveries that cause us to revise what was once thought to be true. A clear example of this was the excavation of Troy in 1870. The city had previously been thought of as a myth based on the work of Homer. The archaeological investigation of this area in Turkey showed that it was a real place and more than just a fiction.

In both cases, we see that 'proof' is a pretty flexible idea – in a way, you could say that it establishes 'the most accurate idea until we know better'. It might not be comprehensive, but evidence is certainly better than superstition – and we have achieved amazing things as a species by using this as our base.

But mathematical proof goes far deeper than its scientific and historical brethren. While science relies on experiments and history uses sources, mathematics has a different tool: logic. This sets it apart in some really important ways. Firstly, it means that anyone can make a mathematical proof. These days, scientific advances are more or less exclusively made by huge teams in expensive laboratories – and this is not something that can be accessed by just anyone on the planet. Mathematical discoveries, on the other hand, can be made by anyone willing to set their mind to the task – often, all you need is a pencil and some paper.

Secondly, mathematical proofs are uniquely durable. Scientific theories are revised over time to reflect better and more accurate experiments, but mathematical truths are timeless. This is why some of the oldest names that we study in school – like Pythagoras and Euclid – were mathematicians, because the theorems they articulated are as true now as they were the day they were discovered. Once we prove that something is true mathematically, it is true forever.

Thirdly, mathematical proofs are general. What I mean by that is that they are true in a wide range of circumstances because logic can help us see why something is not just true in one situation, but in every situation that fits a similar description. For instance, Pythagoras' Theorem states something about the relationship between the sides of a right-angled triangle. It is not just true of the right-angled triangles that we happen to have run experiments on. It is true of every right-angled triangle that can possibly exist.

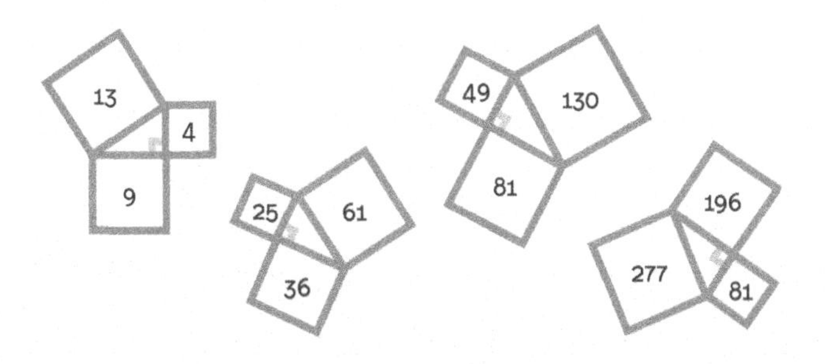

This is really important when we consider the Party Problem, because there is an infinite number of potential parties. It seems like quite a stretch to say that there's a way to know that six is the exact number needed to guarantee that there will be either three mutual friends or three mutual strangers in a party, given the unthinkably large number of possible parties that could be formed from all the people living on the planet.

But this is where the power of mathematical logic is at its best. Walk with me through the argument in the next chapter and see if you can follow along with how it works.

CHAPTER 20

SIX'S A CROWD

'All rise!'

Everyone had been whispering among themselves about who they thought was guilty, but the marshal's call pierces through the conversations and suddenly everyone falls silent. You attempt to read the faces of the other people with you in the gallery, but they are all looking elsewhere – trying to see which door the accused will enter the court from.

As you stand, the room begins to fill with a cast of new bodies. The judge and jury take their positions, and you see a nervous lawyer preparing some notes at his desk near the front. But when you try to find the defendant, you notice something unusual. Instead of a single defendant, there is actually a whole line of defendants, each with their own lawyer, stretching along the side of the room and out the courthouse door. Each is wearing a bright orange shirt with a bold black number written on their back.

The judge addresses the room. 'Today, we are here to get to the bottom of this case. We are here to settle, once and for all, which of you is guilty. Which of you is the smallest number for a party that guarantees at least three people are mutual friends or three people are mutual strangers?' He looks intently at everyone in the line.

'Members of the jury,' he continues, 'it will be your task to weigh the evidence and make a determination. We have a long list of defendants here – every counting number in existence will stand trial if they must! Which number is the smallest, such that a party with that number of people will contain the three mutual friends or strangers?'

The defendants trade nervous glances as they stand in line, which is when you first notice that they are actually standing in order. The defendant with a '1' on his shirt is standing at the front of the line closest to the judge, followed by 2, 3, 4 and the rest of the numbers. You peer out the window and see that the line actually stretches down the street and off into the distance as far as you can see.

One of the lawyers steps forward and addresses the judge. 'Your honour, today I represent my clients – One and Two – and if it pleases the court, I would like to swiftly put to rest the idea that either of them could be guilty of this charge.'

'Go on,' the judge says, nodding. The barrister clears her throat. 'If today's case concerns parties with at least three friends or three strangers, your honour, then clearly you would require a minimum of three people to be friends or strangers in the first place. Neither of my clients fits this description, so I move to have them declared innocent.'

You can hear murmurs of quiet agreement from the jury.

'Very well,' the judge admits. 'One and Two, you have been proven by clear mathematical logic to be innocent. You may leave.' The first two defendants quickly embrace and walk out with their lawyer.

The next person to stand is the lawyer you saw earlier straightening his papers. He looks young, as if he hasn't seen all that many cases, but he has a steely look in his eyes, like he means business. 'That must be the prosecutor,' you think to yourself.

'We are not fazed by this development, your honour, as we have a number of other likely suspects,' he says, without looking up from his notes. Then he shifts his gaze to the next defendant in line. 'Please see Exhibit A of the evidence to convict Three of this crime.'

The marshal brings over a large cardboard sign and mounts it on an easel at the front of the courtroom.

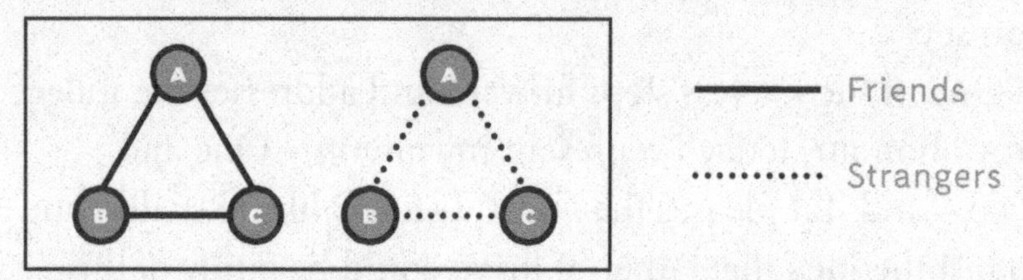

'As you can see,' the prosecutor continues, 'These examples demonstrate the ways in which Three is clearly capable of committing this crime. She has motive and she has the means – what more do you need to convince you?' You can see Three shifting uneasily in her chair.

'Objection!' Three's lawyer answers. 'This is merely speculation from the prosecution. Just because someone is capable of holding a gun, that does not prove that they are a murderer. Just because my client is capable of hosting a party with three mutual friends or three mutual strangers, that does not prove that she has committed the crime. One can hardly say this is proof of my client's guilt, your honour.' He turns to the prosecutor. 'All the prosecution has shown us is conjecture.'

The judge rubs his chin slowly. 'Sustained – you're going to have to do better than that, counsellor. We need better evidence to establish proof.' Before the prosecutor can respond, though, Three's lawyer speaks up again.

'If I may, your honour – I have my own evidence that proves my client's innocence beyond the shadow of a doubt. Could we please present Exhibit B.'

The marshal places a new sign up on the easel:

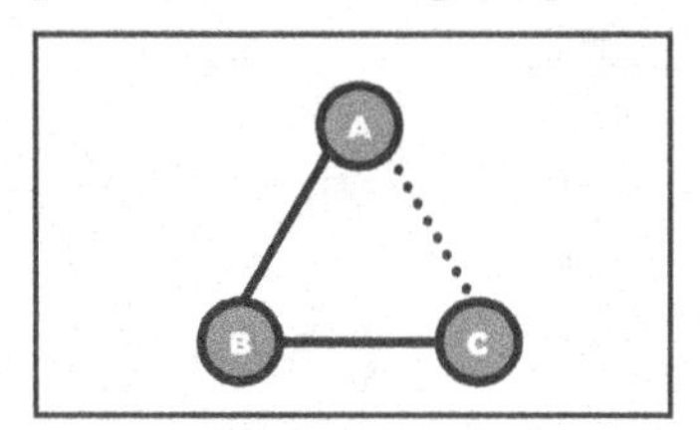

The defence lawyer waits for a moment, giving the jury some time to digest what they're seeing. 'The prosecution has failed to truly understand the charge, your honour. They have not paid attention to one word in particular: "guarantee".'

Three's lawyer goes on. 'My client is not guilty, because a party of three does not *guarantee* that three people will be mutual strangers or three will be mutual friends. The example you see before you now contradicts this statement, because there are in fact no mutual friends or mutual strangers in this party. So much for the prosecution's so-called guarantee, your honour.'

'This clearly contradicts the case you've put forward, counsel.' The judge peers over his glasses at the prosecuting lawyer, who is looking a little embarrassed. 'The defendant is dismissed.'

You watch Three's lawyer as he leads his client away from the courtroom. A new lawyer moves forward to take his place, and she has not one but two defendants beside her.

'I represent both Four and Five,' she begins. 'And if it pleases the court, I would like to call immediately for the marshal to display Exhibit C before the prosecution wastes any more of the court's time.'

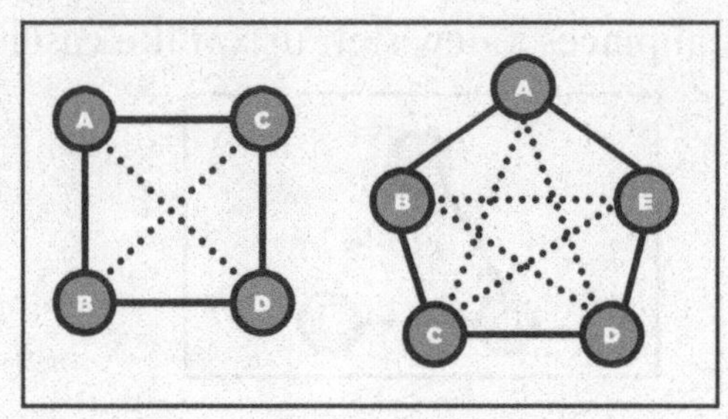

When Exhibit C has been installed on the easel, the lawyer continues. 'As you can see, members of the jury, my clients are also innocent – these images show clear counter-examples without three mutual friends or three mutual strangers. Just as with Three, we have contradicted the prosecution's case and it is incumbent on you to release these defendants immediately!'

The judge nods and looks towards the prosecuting lawyer. 'This is not looking good for you, counsel. Do you have a legitimate case against any of these defendants?'

You can't help but feel a pang of sympathy for the prosecutor. His hair, which had been immaculately arranged when he first entered the court, is now a dishevelled mess and he is looking quite defeated at his post. His desk is a hodgepodge of mixed-up papers and scrunched-up notes, failed attempts to cobble together proof in response to the string of defence lawyers who have triumphantly marched in and out of the court with their innocent clients in tow.

As the judge's last words ring out, though, you catch something out of the corner of your eye. It's the newest defendant, Six, walking slowly to the front of the courtroom. Six looks different from the other defendants who've been on trial this morning. His face is flushed red and beads of sweat are rolling down his forehead. He sits down uneasily beside his defence lawyer, whose briefcase is bursting at the seams with papers covered in hand-drawn diagrams.

The prosecutor straightens in his chair. He's noticed how nervous this defendant looks, and you catch a glimmer of hope in his eyes. He starts to put his desk in order, pulling out a few key pages and writing brief notes on them before arranging the rest of his paper into neat piles. He pauses for a moment, looking at the case he has assembled on the table before him. Taking a deep breath, he addresses the judge.

'Your honour, I realise given my track record today that this may seem implausible – but I believe I have incontrovertible evidence for the guilt of the defendant standing before us.' He continues, this time looking directly at Six. 'I have considered all the possibilities and would like to present my arguments to the jury.'

'Go on.' The judge nods.

'I recognise that this court has a high standard of proof – so to ensure that I am able to prove this to you beyond the shadow of a doubt, I will require you to think in detail about specific relationships in this situation rather than in general terms,' he says, as the marshal places Exhibit D up on the easel. 'It is worth noting that with six people in a party, each person will have a different connection with the others at the party: there are 15 possible connections between each of the people in attendance.'

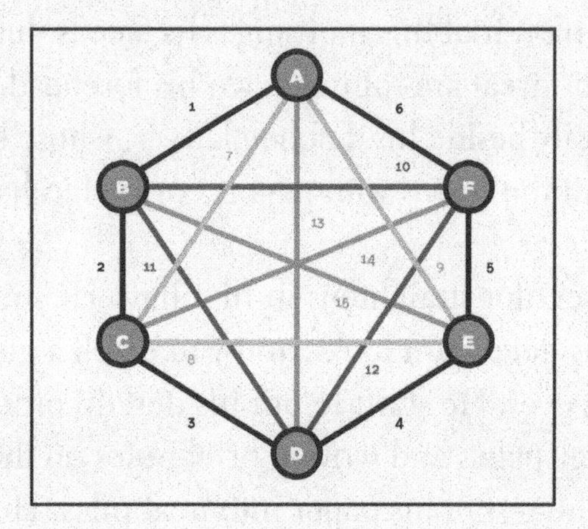

'Each person may or may not have met any of the others, meaning that every single one of those 15 connections will have two different possible values: stranger or friend. This means that for the defendant, Six, there are 2^{15} possible configurations for whether the friends know each other or not. I trust that the court has neither the will nor the inclination to sit through as I demonstrate one by one that all 32,768 of those possibilities result in the defendant's guilt?'

The looks on the faces of the jury give a pretty clear answer to that question. It seems they are just as eager to be out of this courtroom as the defendants.

'I submit, however, that we do not need to inspect every single relationship in every single configuration to demonstrate the required proof. In fact, we only need to focus on a very small network of relationships within the party.'

The prosecutor turns to address the jury directly, as the marshal installs Exhibit E. 'Let's focus on a single person at the party – whom we can call A. There are five other people at the party, and A will either be a friend or a stranger to each of them. As a consequence, the maximum number of friends A can have is five, and the minimum number of friends she has is zero.'

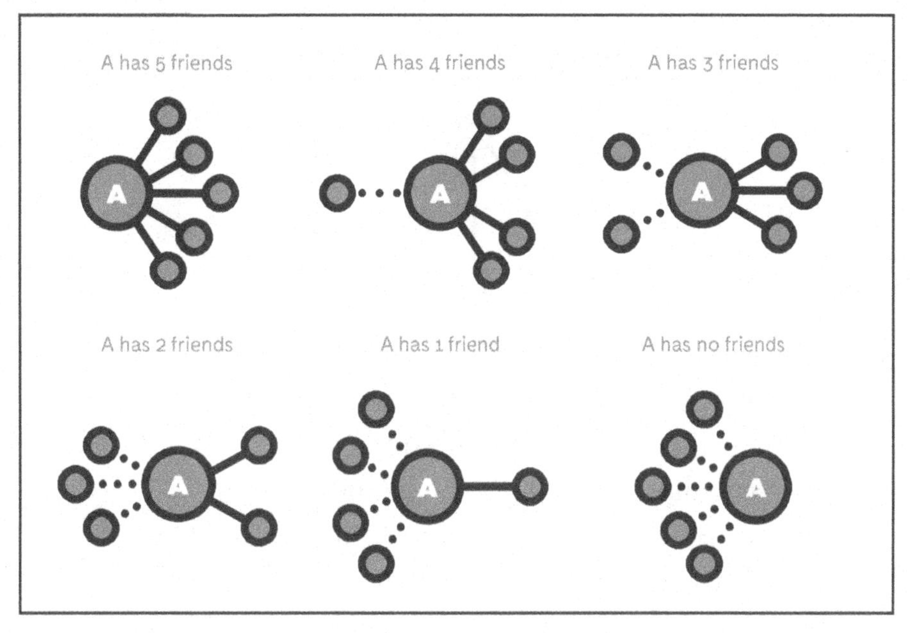

'Please note, your honour, and members of the jury, that no matter which situation we find ourselves in, A will have the same kind of relationship with a minimum of three people. A must have at least three friends, or have at least three people that she does not know – there is no way to avoid this.'

'We are following so far, counsel,' the judge says, nodding slowly, 'but this doesn't prove very much at all about Six's guilt.'

'Quite right, your honour,' the lawyer goes on. 'But let us now focus on these three people who are either friends or strangers to A. It does not matter whether we consider one case or the other, because all the logic that follows can be demonstrated in either circumstance. Suppose these three individuals are all friends with A. Let's call them B, C and D.'

Another exhibit is placed on the easel for the jury's consideration.

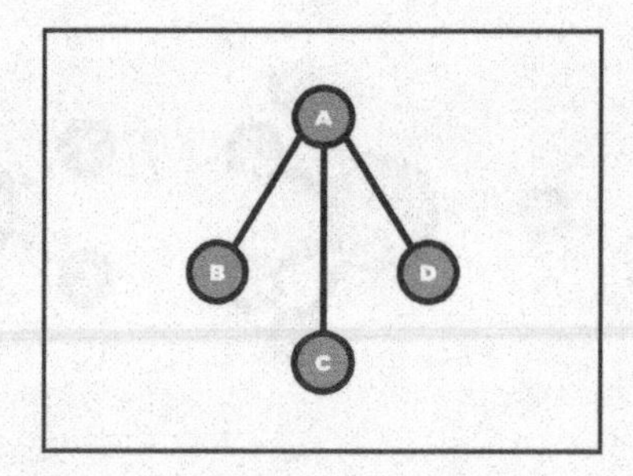

'I submit to the court that if Six is maintaining his innocence, then C cannot be friends with either B or D.'

'And why should that be?' asks the judge.

The prosecutor doesn't miss a beat. 'If C is friends with B, then A, B and C will be mutual friends. Likewise if C is friends with D, then A, C and D will be mutual friends. Three mutual friends – this is precisely the situation that Six is attempting to avoid, your honour.'

You notice Six squirming in his seat a little bit.

'Go on,' says the judge.

The marshal produces another exhibit.

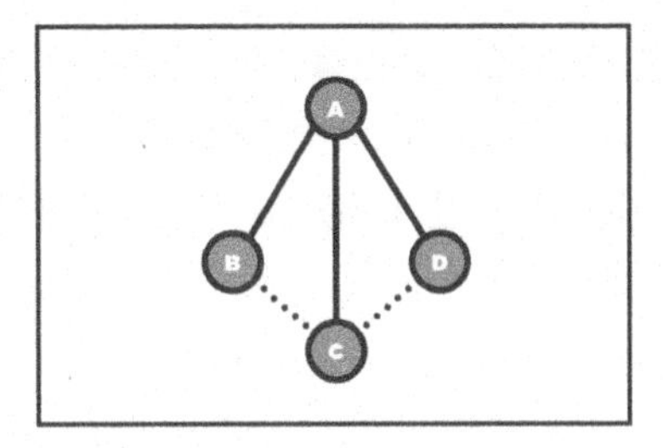

'If C knows neither B nor D, then there is only one relationship left to resolve: the relationship between B and D themselves. It is the final missing link. But your honour, no matter what their relationship happens to be – if they are friends, or even if they are strangers – then Six's guilt is clear.'

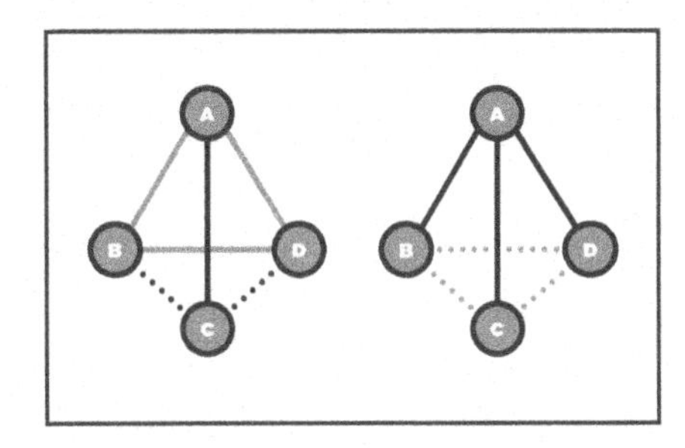

The judge raises an eyebrow. 'How can you be so sure?'

'Your honour,' the prosecutor says with a smile, as he directs the marshal to display one last exhibit, 'if B and D are friends, then A, B and D are mutual friends. If B and D are strangers, then B, C and D are mutual strangers. No matter what happens, a group of three mutual friends or strangers is guaranteed. Your honour, Six is guilty!'

The defence lawyer looks utterly defeated and doesn't even bother to raise an objection. Meanwhile, you've been so absorbed in the prosecution's arguments that you've forgotten to take notice of the jury this whole time. They are nodding in unison to each other and it looks like they will hardly need to deliberate before declaring their decision. The evidence they've seen has convinced them: Six is the number they are looking for.

The judge can sense this too and he bangs his gavel as murmurs start to grow in volume in the courtroom. 'Order, order!' he commands. 'Based on the evidence presented by the prosecution, this court finds Six guilty: six is the minimum number of people required at a party to guarantee either three mutual friends or three mutual strangers. The rest of the defendants are dismissed!'

CHAPTER 21

MY PHONE, THE LIAR

For years, I was convinced that my phone had a vendetta against me. It had a unique ability to get me into trouble through its confident lies. One moment, it would declare to me in strident tones, 'Sure thing, I have plenty of battery life left in me!' – only to die minutes later at a crucial moment, say, during my son's birthday party where I was the trusted photographer.

'What do you mean, you missed him blowing out the candles?!'

'I swear the phone said it had 20% of battery left!'

Other times, the reverse seemed to happen. My phone would appear to cling on for dear life at a mere 2% for what felt like hours. What was going on? Was it trying to make up for all those times when it had let me down?

There's some fairly sophisticated mathematics going into that percentage at the top right-hand corner of your phone's screen next to the tiny battery icon. Yet despite this, it can still get things drastically wrong. To understand why, we need to think a little about what battery life actually is and how it's measured.

Batteries can be loaded up with an amount of electrical charge according to their capacity. Without even realising it, we are conjuring up a physical metaphor when we describe batteries in this way: we are thinking of batteries like jugs, which have more or less capacity to store water depending on their size. When we say that batteries have 'capacity', we are basically visualising them as jugs for electricity.

The analogy breaks down when we realise that unlike jugs, which we can mark with simple lines to indicate water levels, batteries cannot be measured so simply. They are something like an opaque barrel that you can't see inside. How do you measure the quantity of a substance you can't see? We use one of the biggest guns in the mathematical arsenal: calculus.

The mere word 'calculus' is enough to send shivers down the spines of students and adults alike. It conjures up images of mysterious ideas and laws that are inscrutable to normal human minds – literally in some cases. The first time I was introduced to the word calculus was through Professor Cuthbert Calculus, the archetypal mad scientist from the Tintin comics. The professor physically and personally embodied the popular notion of calculus: strange, nonsensical and beyond the comprehension of regular people.

A BRIEF NOTE: Since mathematical ideas are arguably part of the universe itself, they have often been discovered and re-discovered by many people independently around the world and throughout history. The 'invention' of calculus is a particularly contentious issue – the mathematicians Gottfried Leibniz and Isaac Newton got into quite an argument back in the seventeenth century over who came up with it first. Even now, after all these years, people still can't come to a full consensus on who should be able to stake their claim!

But calculus was invented to help us understand a profoundly simple problem: how do quantities change? In particular, when two changing quantities are connected in some way – say, the distance travelled by a car and the time taken to move that far – what is the rate at which they are changing in relation to each other? For instance, how many kilometres (distance) does a car travel per hour (time)?

The direct answer to a question like this can probably come to mind fairly rapidly without much conscious thought. 'Sixty kilometres per hour,' for instance. But taking some time to think at a more abstract level about this idea of a rate – like distance compared to time, which is such an important rate it gets its own name (speed) – may help you to understand some of the most widely misunderstood pieces of language in all of mathematics. Many thousands of people have heard of and even memorised these words and symbols without any real comprehension of what they mean.

For starters, it helps to realise that many common words correspond to mathematical operations. 'And' often indicates addition (for example, 'Can you fetch me five spoons and three forks?' would result in me bringing you 3 + 5 = 8 objects). 'Times' is about multiplication ('I scored a three-point goal seven times tonight' would mean that you scored a total of 3 x 7 = 21 points – nice work!). Though it's less often recognised, 'per' generally means division ('Twelve dollars per pack of four drinks' would mean that each drink costs 12 ÷ 4 = $3).

Therefore, 'kilometres per hour' is another way of saying 'kilometres divided by hours' – which is just 'change in distance divided by change in time'. Since mathematicians are comparing changes in various quantities all the time, they abbreviate it with the letter 'd' (which connects it to the Greek letter delta, the symbol used most frequently in mathematics and science to indicate change of some kind). So *d* (distance) is essentially shorthand for 'the change in distance'.

Since mathematicians love to abbreviate everything (they are constantly searching for more efficient ways of doing things) – they further shorten quantities like time and distance into single letters, called pronumerals. Pronumerals strike fear into the hearts of many students (and adults) around the world – mainly because we've never understood what they're about. To put it simply, pronumerals are just like pronouns:

* Pronouns are words (he, she, it) that stand in place of nouns (man, woman, book).
* Pronumerals are symbols (x, y, a, ø) that stand in place of numerals.

The most common pronumerals that we encounter in mathematics are x and y, so if we write x instead of 'time' and y instead of 'distance', then we can abbreviate *d* (distance) / *d* (time) as dy/dx. These four letters are like the 'abracadabra' of calculus. Hundreds of thousands of students write these letters over and over again each day around the world without grasping what they actually mean. They aren't a magical incantation – they're just an abbreviation for expressing how two things change in relationship to each other.

One last note here before we return to batteries – when we are comparing two changing quantities, we often like to draw them onto a graph or chart. These graphs can compare anything – for instance, a weather chart might show maximum daily temperatures in comparison with the time of year. Or we might have a graph that shows a company's total revenue on an annual basis. When my wife and I had our first child, the hospital gave us a little book that included 'growth charts' – this compared a child's age with their expected weight and height for that stage of development.

When we draw such graphs, it's conventional to label the horizontal axis as x and the vertical axis as y. This means that when we think about the 'change in y', we're trying to work out how much vertical change occurs: an easy way to describe this is 'rise' (i.e., how far the graph goes up). On the other hand, when we think about the 'change in x', we're looking instead at how much horizontal change occurs: people have taken to calling this 'run' (i.e., how far the graph goes across). The dy / dx we mentioned earlier, then, is remembered by many as 'rise over run'.

We've just covered a huge range of ways to describe the same idea – how something is changing over time – and that diversity is owing to the fact that rates of change exist everywhere around us and we're consistently interested in studying them. Here's a visualisation of the language path we just strolled down:

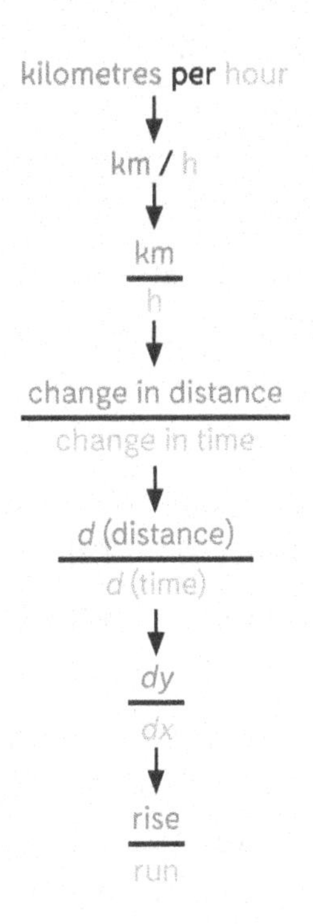

These are the cogs and wheels of the great mathematical machine that we call calculus.

What does all this have to do with phone batteries?

The initial problem we were wrestling with was trying to find a way to measure a quantity you can't see: namely, the amount of charge in an electrical battery. If we imagine a battery as a jug, then it's one that you can't see inside. So how can we work out how full it is?

A metaphor will help here. When I was at school, I remember going on overnight camps where we stayed in cabins, each of which housed ten to 12 students. Due to the simple facilities, we all had to take it in turns to use the showers. Each night, after the day's activities had come to an end, there would be a mad rush to get to the bathrooms. Friendships and loyalties meant nothing at shower time. All that mattered was getting to the bathroom first and ensuring you got to shower sooner rather than later.

Why? Well, there was usually a limited supply of hot water on the camp site – fed to each of the cabin's bathrooms from a large tank – and as the hot water level dwindled, the temperature and pressure of your shower slowly dropped accordingly. Those blessed with quick feet (or who had been allowed to get to the showers more quickly by their supervising teachers) could enjoy a blasting hot shower, while those who got there too late had to put up with cold showers where the water barely trickled out of the faucet. Camp trained us – either to be fast or develop resilience (especially during winter camps)!

Batteries are a little bit like the water tanks on those camp sites. The fuller they are, the faster their flow – of hot water in the case of the tanks, and electrical current in the case of the batteries. So while it's impossible to directly know how much charge is left in a battery at any given time, it is possible to get a rough idea by measuring the rate at which electricity is being dished out by the battery. Since this then becomes a question about rates of change, this is where calculus is in its element.

Phone manufacturers do lots of laboratory testing on their batteries so that they can predict the amount of charge that corresponds to a particular flow rate of electricity. They calibrate the devices they produce to recognise these rates and report a battery level accordingly. So what's the big problem? If it's so straightforward to work out how much charge is left, why do our phones get it so wrong much of the time?

There are a bunch of factors at play here. For starters, batteries don't operate at the same efficiency in all conditions. In very high or very low temperatures, batteries find themselves unable to hold on to their charge for longer periods of time, so they don't last as long outside of their optimal operating temperature. Not only that, but batteries don't drain at a uniform rate that can be easily predicted. Just like the water in the tank is depleted faster if there are lots of people having hotter showers, different functions (like cellular data) or apps (like video editing software) on your phone require more processing power and so they chew through the battery more quickly. Finally, batteries actually lose their ability to hold charge as they age. If you feel like your old faithful phone doesn't seem to last as long as it used to, then you aren't imagining things!

The software in your phone and its mathematical algorithms for determining battery charge will do their best to cope with these changing circumstances, but ultimately all they are providing is the best estimate they can. That percentage up in the top right corner, which seems through its numerical precision to suggest an exact knowledge of how much battery life you actually have left? No joke – it's lying to you.

A BRIEF NOTE: It's not lying intentionally, of course. Your phone's way of calculating battery life uses a mathematical model, and as the British statistician George Box said, 'all models are wrong – but some are useful', and arguably this one is!

CHAPTER 22

MATHEMAGIC

I love a good magic trick. Good magic tricks don't so much fool me as they make me question my own ability to know anything with confidence. I've been lucky enough to see some great magicians in person and each time it's happened I've come away literally not believing my own eyes.

Many tricks require special and expensive props or years of training in misdirection and sleight of hand. But tricks like the one I'm about to show you only require a deck of cards. That's because they work not through equipment or manipulating perceptions, but by mathematics.

Take a full deck of cards – 52 cards across the deck, with the jokers removed – and give it a good shuffle (if you are showing someone else this trick, have them shuffle the deck). Then divide the deck into four piles according to the following steps:

1. Uncover the first card in the deck. If it is red, place it face-up on your left. If it is black, place it face-up on your right.
2. After you have placed the first card, take the second card – which you have not uncovered or looked at – and place it in its own pile above the first card. You now have your first two piles: an uncovered pile and a secret pile.
3. Uncover the third card, and treat it just like the first card – place it face-up on your right or left in the appropriate red or black pile.
4. Take the fourth card and treat it just like the second card – without looking at it, place it face-down in its own pile above the third card.

Repeat this process until you've gone through the entire deck.

If you've been following so far, you should have four piles of cards in front of you: a red uncovered pile, a black uncovered pile, and two secret piles – one for each colour.

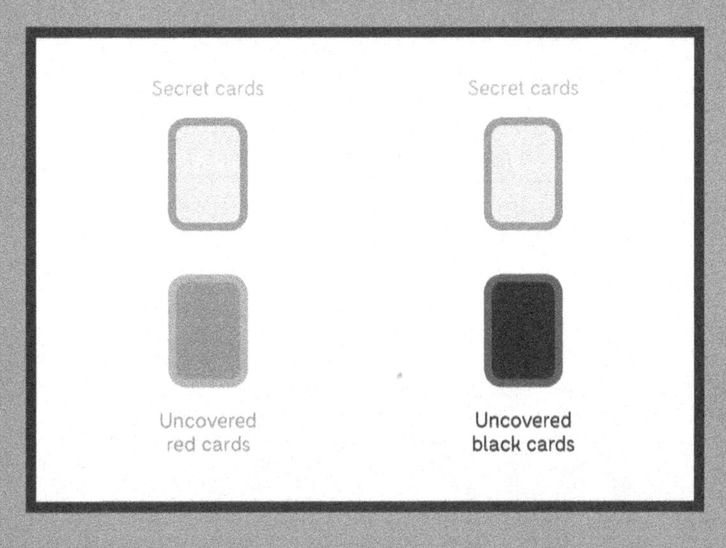

Before you move any further, I want you to consider a couple of things about those secret piles. The two piles are probably different in size, meaning that the number of cards in each pile is different (and it's likely that you don't know how many cards are in each). In addition, the cards in those piles truly are secret: you haven't seen any of them; you've placed them down with no knowledge of what kind of card each one is. I'm just trying to establish that you really don't know very much about the secret cards.

And in a few seconds, if it were possible, you're going to know even less. Again, if you are showing this to a friend, ask them to help you with these steps:

1. Tell them to select a random number between 1 and 6. If you can find a die, use that to make things extra random.
2. Just say that they select the number 5. If that's what they choose, tell them to take out any 5 cards from one secret pile.
3. Now, without looking at any of the cards, swap those 5 cards with any 5 cards from the other secret pile.

Now the secret piles have been jumbled up with each other and there's no way you could know anything at all about the contents of the secret cards.

Or is there? Here comes the trick: tell your audience that you will now make a magical prediction. You predict that the number of red cards in the secret pile on the left is the same as the number of black cards in the secret pile on the right. Go ahead and uncover all the secret cards to check. You'll see that your prediction came true! Do it again and you might get a different number of cards in the secret pile, a different number when you roll the die, and a different number of reds or blacks in each of the secret piles. But you'll always get the same result. No props or special skills required!

So how does the trick work? The easiest way to get to the bottom of what's happening is to unpack an actual example, step by step. First we'll work out what the four piles are doing, then we'll look at the random swap that happens at the end just before the big reveal.

While it's true that there's no sleight of hand involved in this card trick, there is in fact an illusion that I've been trying to disguise. While it may appear that we don't know very much about the secret piles, we actually know a tremendous amount. If we apply a little bit of logic – mathematical logic – to the rules that make the trick work, you can see that we know absolutely everything we need to know to ensure that the trick will work every single time.

The first thing you need to recognise is that there are only two colours in a deck of cards: red and black. That means exactly half of the cards, 26, are red, and 26 are black. Remember that for a moment.

Let's fast-forward through the whole process as if we've just finished dealing the cards into their four piles. As you may have noticed, you will get a different number of uncovered red cards and uncovered black cards each time. But because of the way that you've dealt out the cards, alternating between uncovered cards and secret cards, you can know the following:

* Exactly half of the cards, 26, are uncovered, and exactly half of the cards, 26, are secret.
* The number of cards in the uncovered red pile and the secret pile beside it is the same.
* The number of cards in the uncovered black pile and the secret pile beside it is also the same.

These are all the facts you need. Now all we need to do is a little bit of counting – and watch what unfolds!

Here's an example of what the cards look like after I dealt them out according to the pattern. Notice how many cards there are in each pile and how they fit exactly what we expected by using our logic from before.

This trick revolves around the black and red cards in the secret piles. While we're performing the trick, we don't actually know anything about the cards in each individual pile; but since we're trying to unmask the trick, let's turn over one of the piles and see if we can work out what's happening. Suppose we do that and this is what we observe:

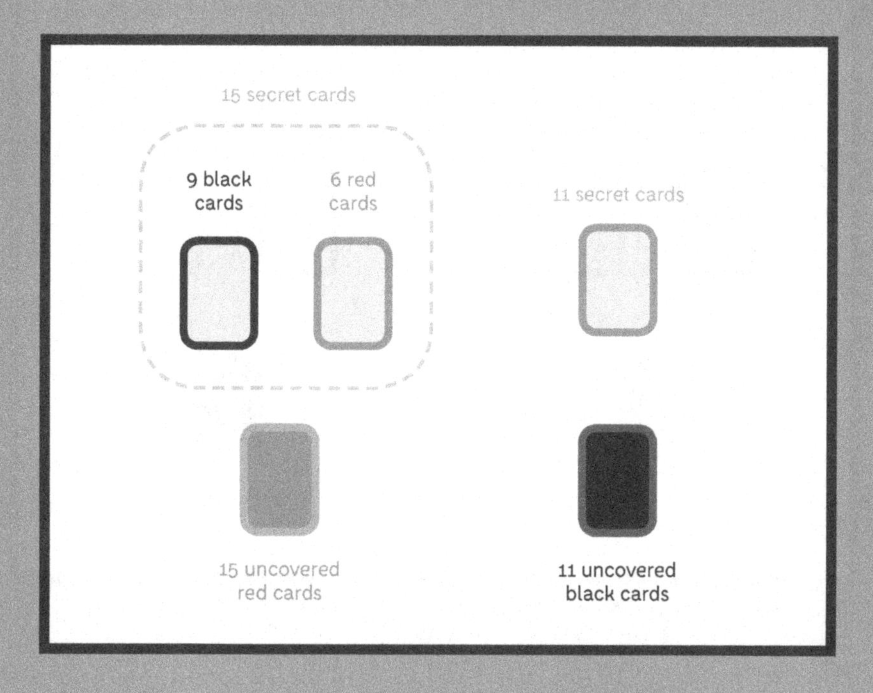

At this point we could uncover the other secret pile and count things out, but this would be the same as just performing the trick. Instead, before uncovering the final pile, let's again use some mathematical logic and see if we can work out what's happening first.

The key piece of information, as we noticed earlier, is that in a standard deck there are exactly 26 red cards and 26 black cards. As you can see, we've already identified where 21 of the red cards are. That means the remaining 5 reds must be in the final pile. But if 5 of those secret cards are red, 6 must be black:

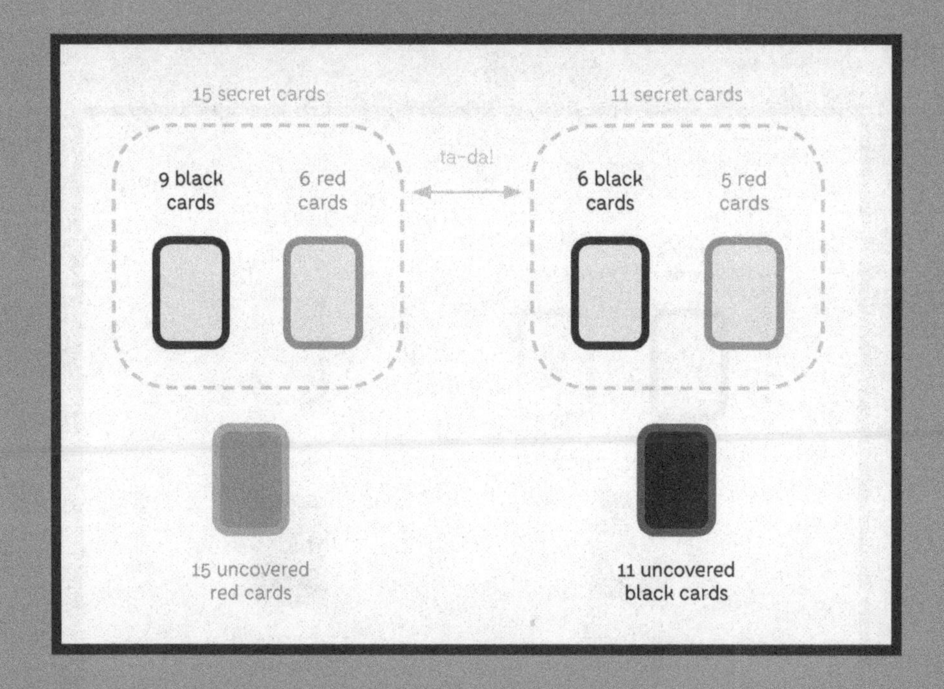

There are 6 reds on the left and 6 blacks on the right; the trick works! And no matter how many different times you do the trick, it still works; you may start off with a different number of red and black uncovered cards, or a different combination of red and black secret cards, and logical counting will show you that it works every time.

In fact, this is one of the reasons why we learn algebra in high school. No, not for dissecting unusual card tricks, but for working with numbers even when we don't know what their values are. Since we don't know what they're equal to, we refer to them with letters called pronumerals, which we discussed in the last chapter.

In the diagram on the opposite page, I actually did the trick and found I had 15 uncovered red cards and 6 secret red cards beside them. But algebra allows me to place a pronumeral in place of 15 and a pronumeral in place of 6, after which I can imagine them to be absolutely anything I like – 12, 8, 23, and so on. If you want to check out my proof of how the card trick works with algebra, you can duck into the next chapter and have a look at my explanation there.

'Wait a second,' you might be thinking. 'What about the weird bit at the end where we swapped some of the cards randomly? You didn't explain that bit!' Good point! First, let me give you a tiny warning – the reason this trick works and successfully surprises people is because it has just enough maths in it to make it a challenge to follow. But if you can watch the cards move around – and I highly encourage you to find a deck of cards to follow me as I walk you through this – then you'll be able to see what's going on and have fun messing with the minds of your friends and family!

To get to the bottom of what's happening, let's forget about those uncovered piles and focus on the secret piles. Just like in the trick, let's actually take out 5 cards at random from each pile and see what happens.

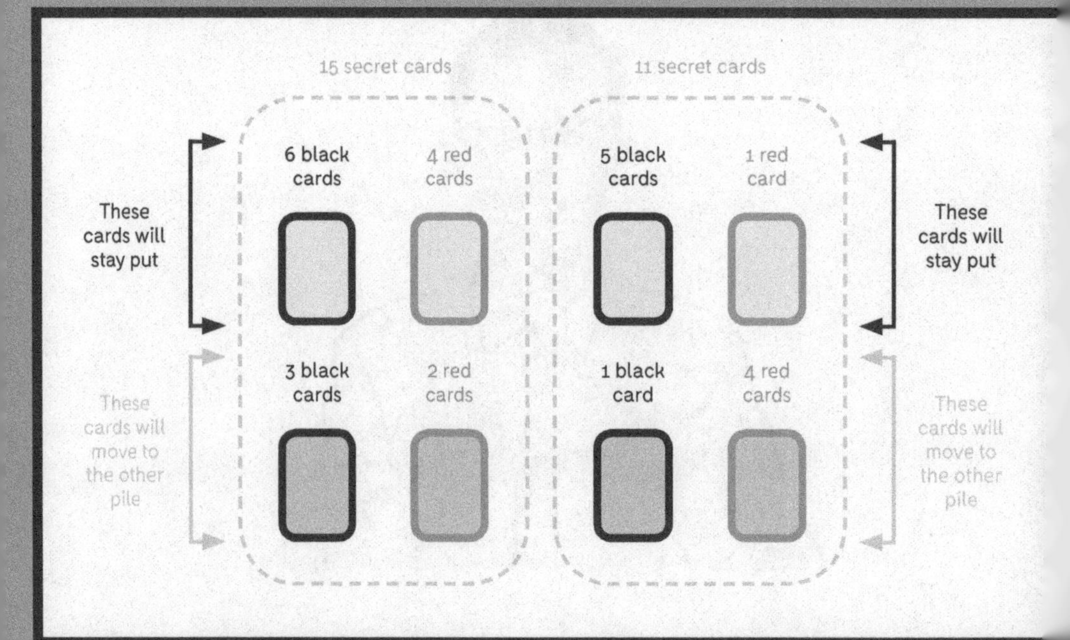

Here you can see I've pulled 5 random cards out of each secret pile and set them aside. I want to note their colour before I do the switch, because that will allow me to keep track of the numbers once I swap things around. Notice that the cards I'm about to switch aren't the same; there's a different combination of reds and blacks that is going to switch places.

So far I've just selected 5 cards to swap, but I haven't moved anything yet; everything on the left is still beside the uncovered red cards, and everything on the right is still beside the uncovered black cards. Now let's do the swap!

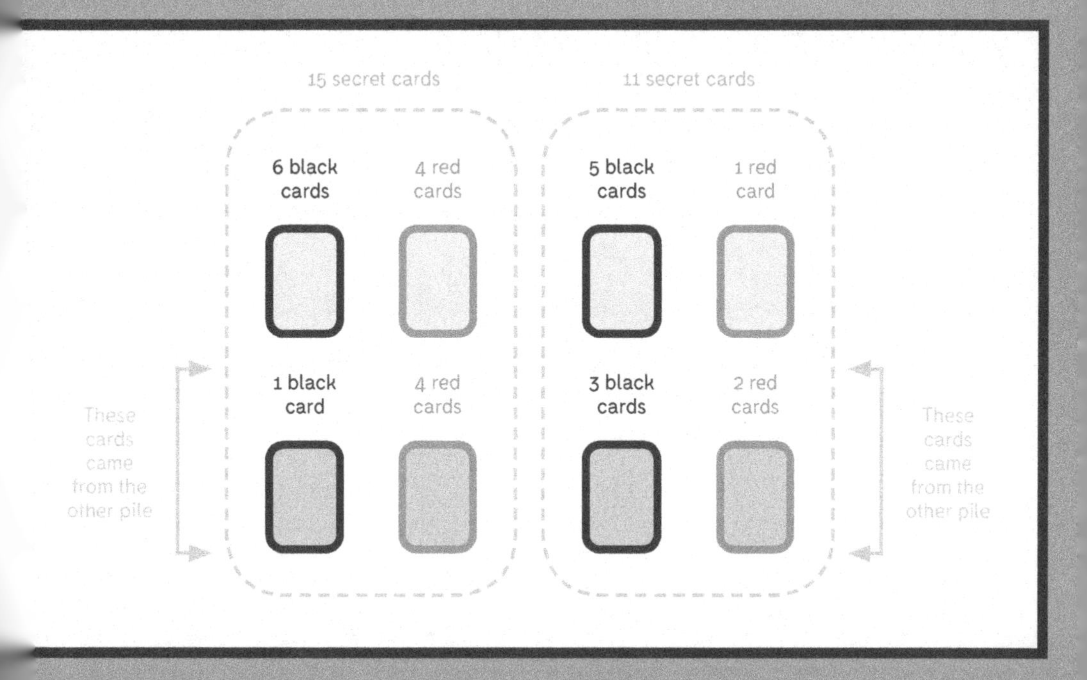

You can now see that the 5 randomly chosen cards are in their new homes. So what happens when I square everything up and find out the new totals for each colour?

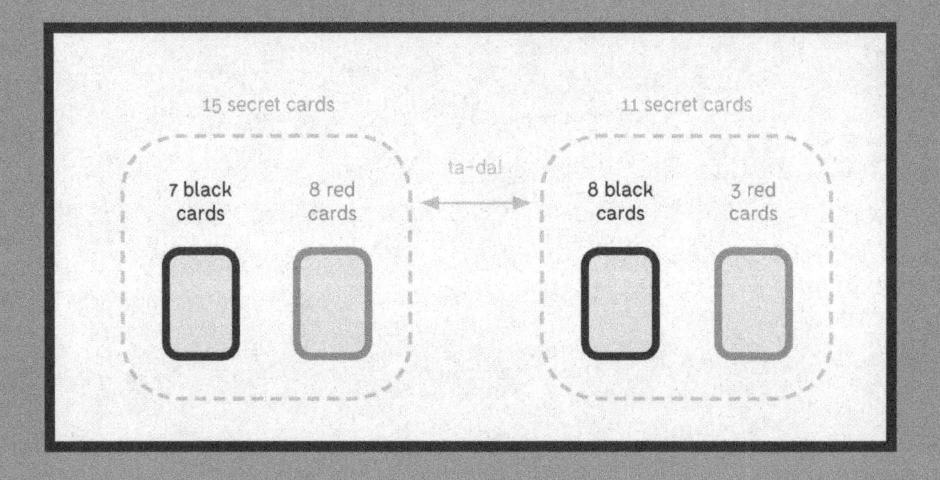

Lo and behold – still the same number of reds on the left as there are blacks on the right!

Why doesn't the swap end up changing anything, even though all our intuition says it should? Again, if we're ready to put on our thinking caps and apply some logic, the reason becomes clear. Allow me to show you.

To help you understand, I'm going to employ that handy technique I mentioned earlier: when a problem is hard to solve, we imagine a simpler version of the problem and see if we can understand how it works. It's easier to understand a simpler problem than a more complicated one, and often the insights you gain from the former can help you with the latter.

So instead of swapping 5 cards between the piles, let's just imagine swapping one card between each pile.

Think about what it would mean if both cards were the same colour, say black. Remember, all we're interested in is the card colours and not their values or suits. That means once the black cards swap places, the result is identical to what we started with. All the counts for reds and blacks will remain as they were before we made the switch.

The same is clearly true if both cards are red instead of black.

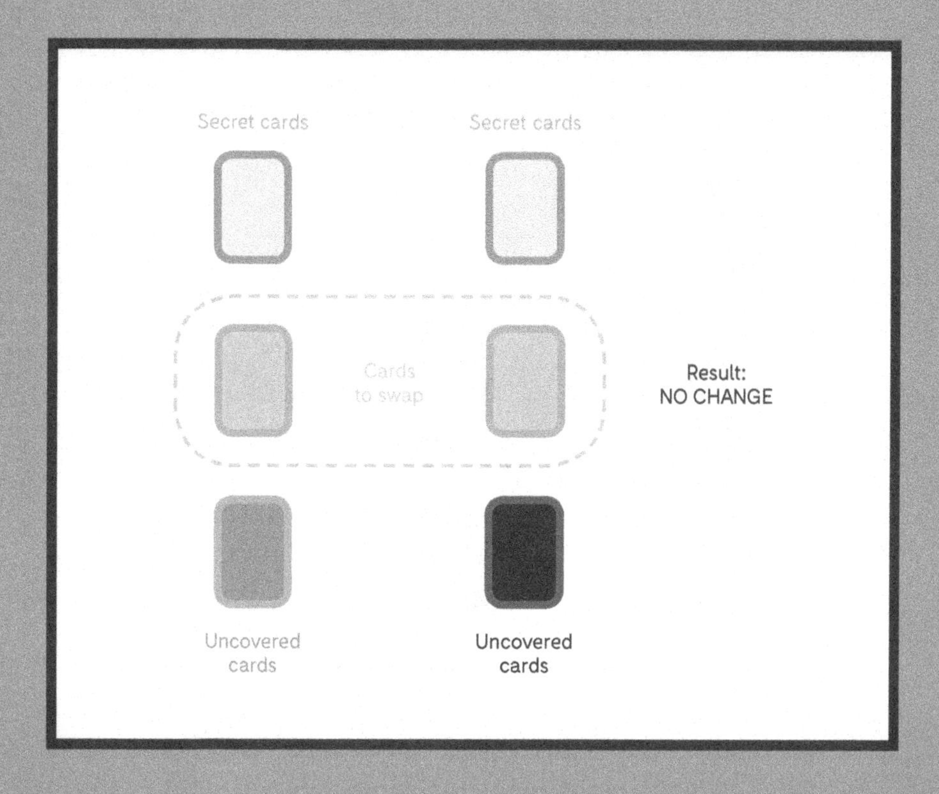

But what if the cards are different colours? That isn't too hard to work out either: if we take a red card from the left and move it to the right, then the number of reds on the left secret pile goes down by one; but we're also taking a black card from the right and moving it to the left, so the number of blacks on the right secret pile also goes down by one. The numbers will change, but the numbers that matter to us – the reds on the left and the blacks on the right – will both decrease by one and stay equal to each other.

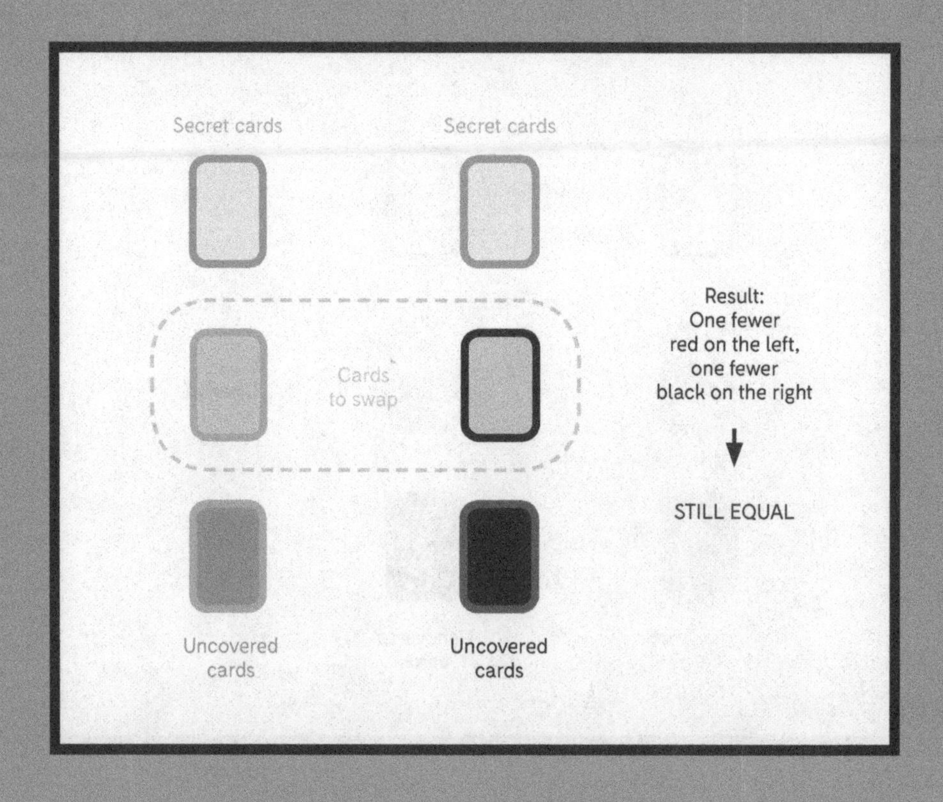

The opposite can happen as well, where the numbers of reds and blacks both increase by one – but still stay equal.

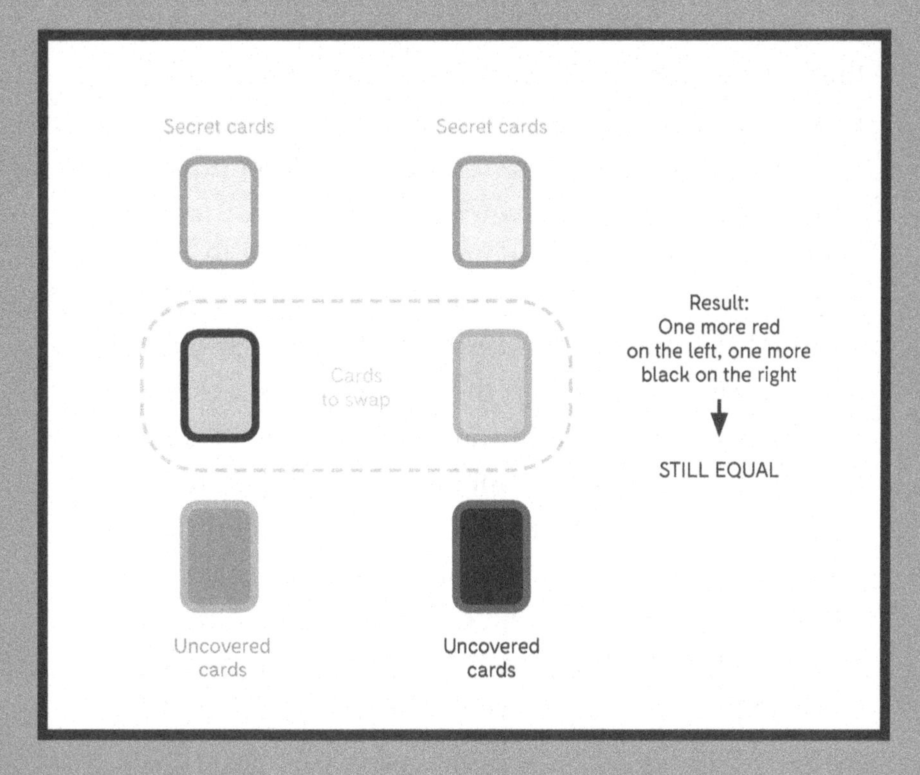

So now we can see that if we swap just a single card from each pile, then every single one of the possibilities will lead to the numbers ending up equal. Here's how we can relate that simple problem to our more complicated version: switching five cards all at once is exactly the same as switching one card five times in a row. If performing the swap keeps the red and black counts identical, then you can do it as many times as you like and they'll always be equal to each other.

So what's the point of this exercise? Well, for starters, it's great fun – I've lost count of the number of people, young and old, who I've entertained and perplexed with this little trick. There's also something charming about the fact that it can still work even if you don't fully grasp how it works. While sometimes it can be frustrating when something works even though we don't understand why, there are times when we rely on that very fact to get by. When I push down on the accelerator in my car it goes forward, even though I'm largely ignorant of the hundreds of moving parts and the complicated physical and chemical interactions that make it work – and when I'm driving down the highway at 110 kilometres per hour, I'm quite happy to not have to worry about the exact mechanics of my car's internal combustion engine! It's a testimony to the power of engineering that cars can work without us intimately understanding them, and it's a display of the inner harmony present in all mathematics that we can wield it even before we fully comprehend it.

But understanding is still a goal worth striving for. True understanding gives you x-ray vision to see into phenomena that are opaque to others – and when you really grasp how something works, you can appreciate it more deeply. Many patterns in our world, both the natural world and the human one, are animated by mathematical principles that are invisible to most of us – just like this clever little trick. That's why, if we can bring ourselves to pause for a moment and think with clarity and logic, sometimes the true nature of a mathematical reality can open itself up to us and become as easy to deal with as a deck of cards.

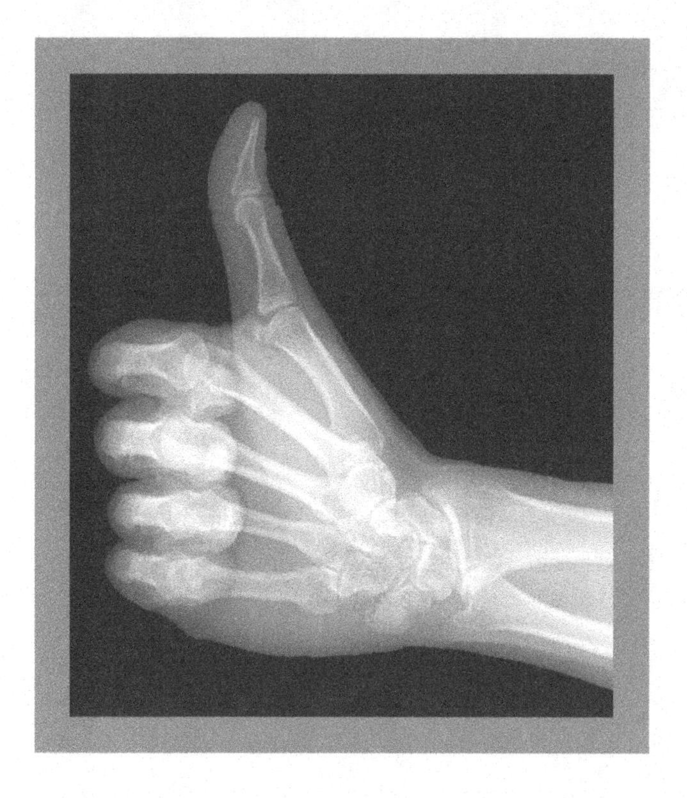

CHAPTER 23

MAGIC WITHOUT MISDIRECTION

In the previous chapter, we used a standard deck of playing cards to illustrate that wonderful mathematical patterns can emerge out of seemingly random processes if there is just a tiny bit of structure or regularity hiding within them. I demonstrated, with a concrete example and actual counting of the cards, why the numbers always work out.

I'm sure some of you want to know, though, why the trick works every time – and why you may get different numbers each time, but the numbers of red and black cards in each secret pile will nonetheless match. In other words, you want to lift up the bonnet and peek inside the engine to see what's going on. I can't blame you! But be warned – to really get to the bottom of this, you're going to need to embrace algebra.

I'm flagging this right up front because I know how often people bristle even at the mention of this part of mathematics. It is one of those things that people remember in the same way I used to think of the big boss characters in computer games I played as a kid. 'Oh yeah, I loved playing Metroid! Until you get to that level where you have to fire a million rockets at that giant dragon thing with a thousand teeth and fifteen eyes. I never got past that boss. It sucked.' When people find out at parties or wedding receptions that I'm a mathematics teacher, this is the way they talk about algebra. 'I didn't mind maths, until they started mixing letters in there!'

But algebra is one of the most powerful tools that human beings have ever developed to help them solve problems. It's because many

> Dear Algebra,
> Please stop asking us
> to find your x.
> She's never coming back,
> and don't ask y.

problems in the world include numbers whose value is either uncertain or variable. And this is where algebra comes in: it says, 'You don't know what that number's supposed to be equal to? No problem. Let's just put a placeholder there for now – a letter will do. You can swap it for the number when you find out what it is, if you want.'

As we mentioned earlier, these mathematical placeholders are called pronumerals. If you think about the card trick from the previous chapter, it is pretty apparent why pronumerals are going to come in handy. How many cards get dealt out face up on the red side? I don't know until I try the trick and count them. And each time I do the trick, the number changes. But that number will define much of the rest of the trick, no matter what it is equal to, so I can substitute it for a pronumeral and work with it anyway. Hold on to your hat and I'll show you how it all works!

Okay, first off, follow the steps from the previous chapter up until you've dealt out all the cards from your deck. (As with last time, we will work out what happens with the card swap later on.) This is how things should look right now:

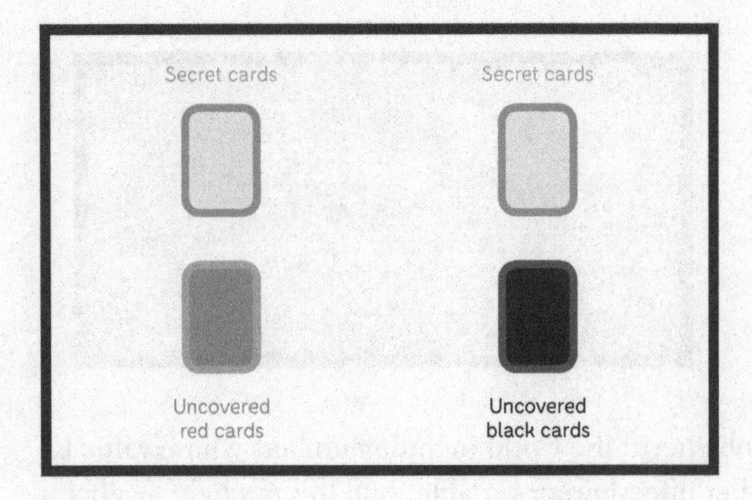

One of the essential principles of mathematics, and especially algebra, is that it loves to speak in succinct terms. An expert in grammar would say that mathematicians love to 'maximise lexical density' – that is, squeeze the most meaning into the smallest space. So rather than use long phrases to describe things, let's label these various piles with concise names to make them easier to refer to. We'll call the left-hand piles R1 and R2 (since they are on the red side), while the right-hand piles can be called B1 and B2 (since they are on the black side).

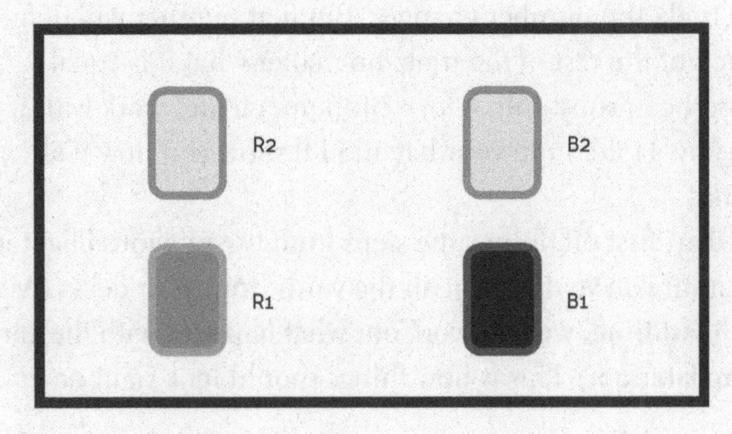

One of the key insights into this situation is to realise that the size of each pile you've made here is closely related to the others, even if this is not obvious on the surface. To illustrate this, let's use our next algebraic technique and give a name to the number of cards in R1. In the worked example I went through last chapter it happened to be 15, but it could be anything – so let's follow an old mathematical tradition and call it x.

Since pronumerals stand in place of numerals, they follow all the same rules as numerals. This means we can apply all the same logic we did in the previous chapter to come to some conclusions about how the piles are related to each other. We drew a few conclusions last time, so let's now state them in terms of the algebra we've introduced:

* 'Exactly half of the cards, 26, are uncovered, and exactly half of the cards, 26, are secret.' If there are x cards in R1, then the number of cards in B1 must add together with these to make a total of 26. This means there are $26 - x$ cards in B1. In the last chapter, these numbers turned out to be 15 in R1 and $(26 - 15 =)$ 11 in B1.
* 'The number of cards in the uncovered red pile and the secret pile beside it is the same.' Every time you placed a card in R1, you also placed a card in R2 – so these piles must end up exactly the same size. There are x cards in each.
* 'The number of cards in the uncovered black pile and the secret pile beside it is also the same.' This follows on from the previous two points put together. Since we've already established that there are $26 - x$ cards in B1, there must be $26 - x$ cards in B2 as well.

All right, let's check in – this is what we know so far:

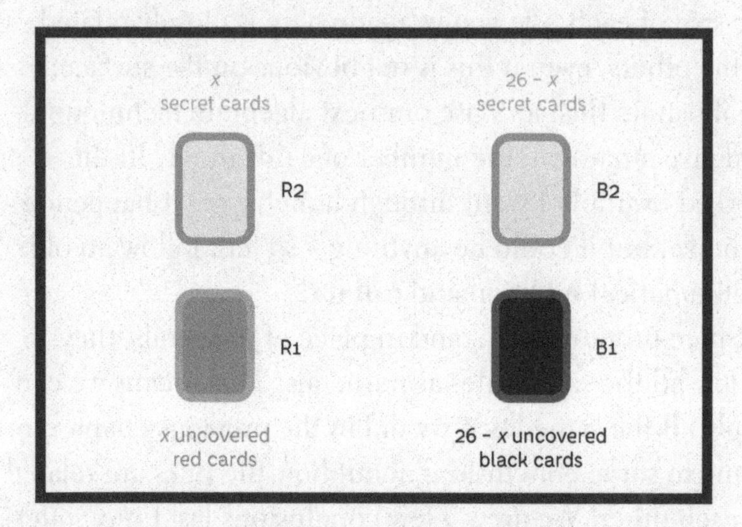

A quick mental check also confirms that if you add up the numbers in all four piles, the total is 52 cards – as it should be!

For the next step, we need to zoom in a little closer to R2. As the trick unfolds, what we're interested in here is the number of red cards hidden in this pile. If you did the trick more than once, you'll notice that just like *x*, this number changes every time. In this case, let's label it *y*.

Now is the time to remember that we actually know, right from the start, how many red cards there are in the entire deck: exactly half of them, which is 26. There are *x* of them in R1, and *y* of them in R2. Remember that there are no red cards in B1 (because we only placed cards in B1 if we saw that they were black). This means that if there are any red cards left, they must all be in B2.

Think about that again for a second – it's a crucial step in the logic and you need to get it firmly in your head. B1 is completely made up of black cards, which means that there are exactly 26 red cards distributed throughout R1, R2 and B2. We already know how many are in R1 and R2. Before I reveal it, can you work out how many red cards must remain in B2?

To balance everything out, there must be exactly $26 - x - y$ red cards hidden in B2. When you add these red cards to the x red cards in R1 and the y red cards in R2, you will come up with the correct 26 red cards that are supposed to be in a standard deck (unless there was a serious malfunction at the card-printing factory!).

Here is where the trick finally falls into place. Do you remember what the trick's prediction was? The number of red cards in R2 is supposed to match the number of black cards in B2. Can algebra help us work out how many black cards are in B2 before we actually count them?

I'm now ready to unleash our final algebraic weapon on this problem – the equation! It is perfect to help us nut out what's happening in B2.

Number of black cards in B2 = (Total number of cards in B2)
- (Number of red cards in B2)
$= (26 - x) - (26 - x - y)$
$= 26 - x - 26 + x + y$

(In case you've forgotten your high school maths, subtracting a negative is equivalent to adding a positive!)

Number of black cards in B2 = y

Number of black cards in B2 = Number of red cards in R2

And there you have it!

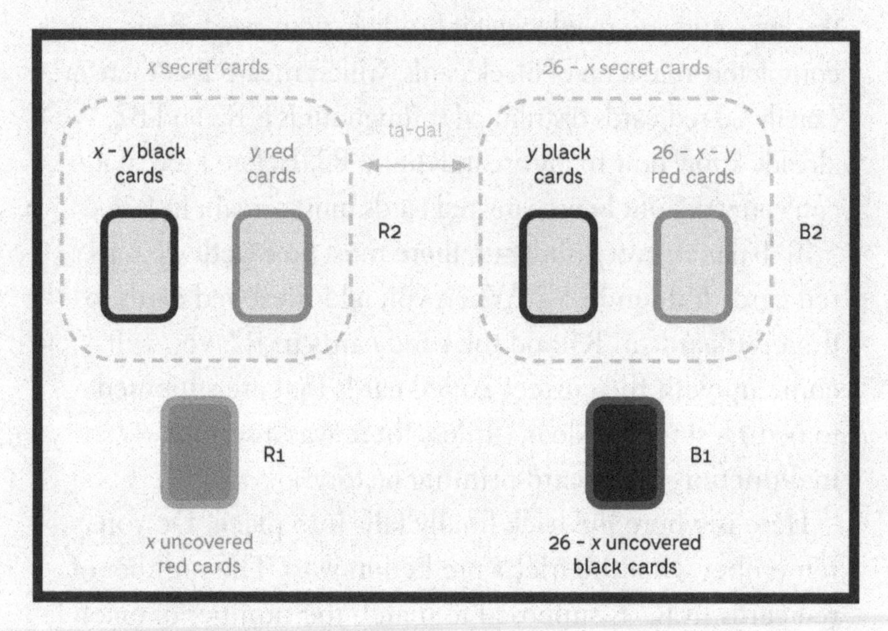

Look familiar? This is the algebraic version of what we saw on page 288.

Right, then – now it's time to tackle this pesky card swap! This is even more challenging to grapple with because, unlike in the first half of the card trick, the audience gets to participate and make a genuine choice by choosing how many cards to swap. However, if you can stick with it, the payoff is even better when you see it all come together!

For this part, we only need to think about the secret piles (R2 and B2), since that is where the swap takes place. The audience gets to pick any number of cards they want to swap between R2 and B2 – let's call this number n.

The audience also gets to control which cards get swapped, and we have no idea if these are red or black –

which means more pronumerals we have to keep in mind. However, we can keep the pronumerals to a minimum by concentrating only on the numbers that really matter to us. We are interested in the red cards from R2, so let's say that the audience chooses a red cards from R2 to move into B2. This leaves behind $y - a$ red cards in R2. Since they are swapping n cards in total, they are going to move $n - a$ black cards over into B2 at the same time.

In the same way, we are interested only in the black cards from B2, so similarly we can say that the audience chooses b black cards from B2 to move into R2. This leaves behind $y - b$ black cards in B2. It also means that $n - b$ red cards will move over to R2 once the swap is done.

So just before the swap happens, this is the state of your cards:

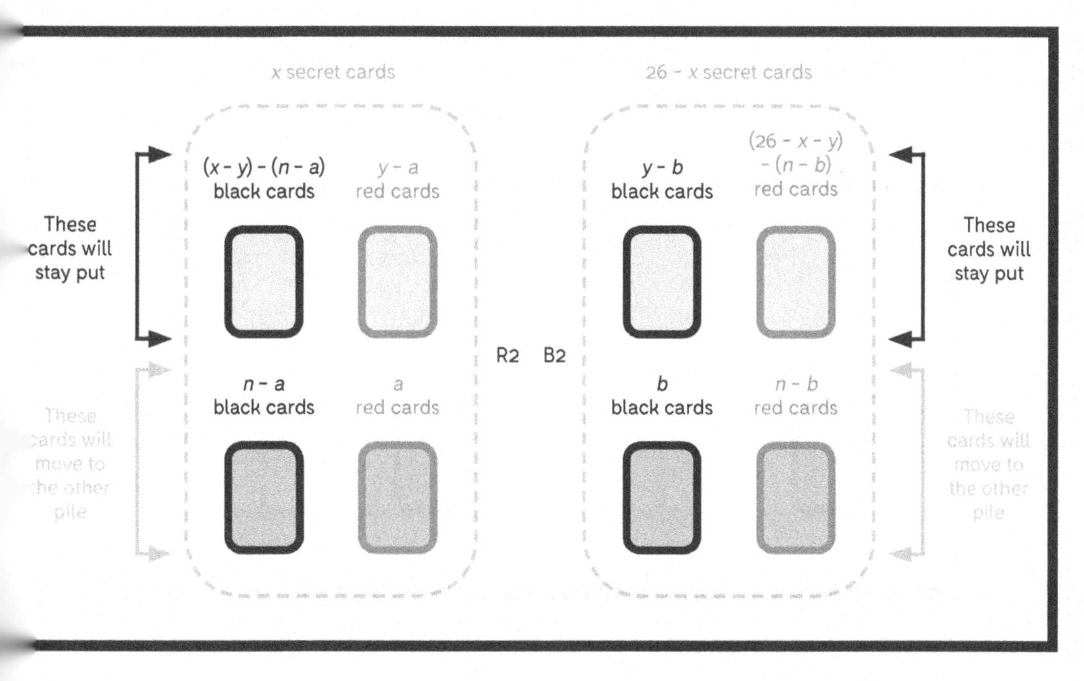

Once the swap is actually done, this is how things look:

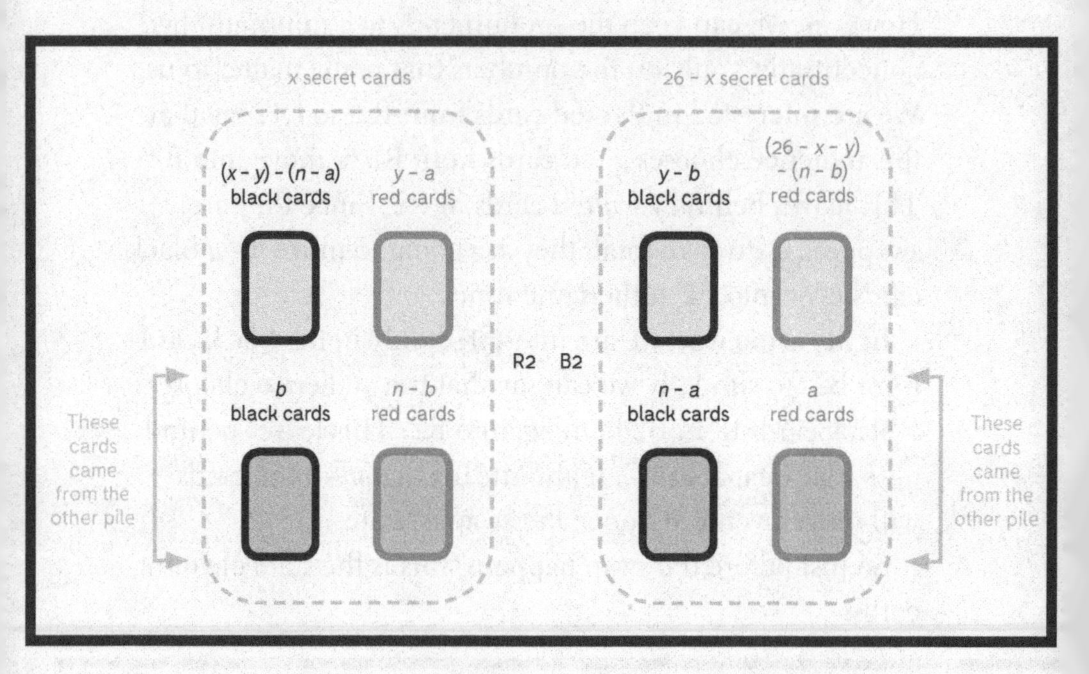

And if we neatly combine the swapped cards with their new homes, this is the final state of the card trick:

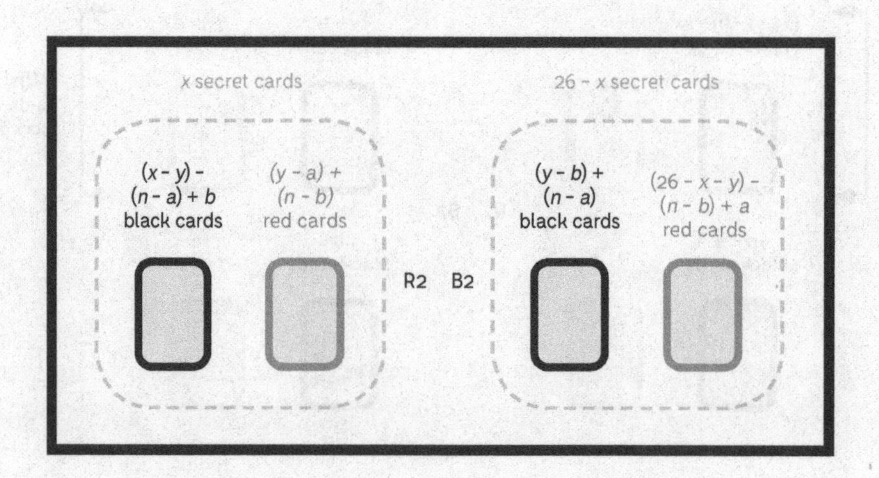

But if you remove the brackets and look closely at the red cards in R2 and the black cards in B2, you'll see that things still marry up beautifully (just as they did back on page 292):

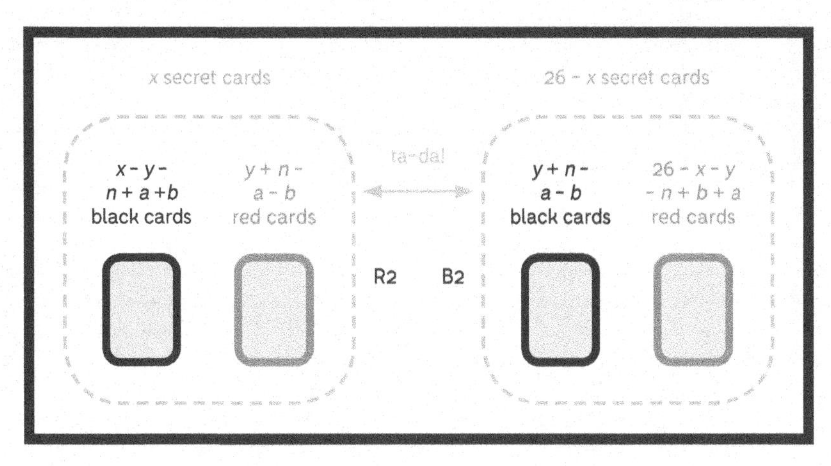

So what's the point? There are plenty of things in our world that seem as though they are random and chaotic, with no rhyme or reason to make sense of them. But in many more cases than we might expect, there are actually patterns and logic hiding beneath the surface. Mathematics is one of the finest tools ever conceived of by the human mind for making sense of the world around us. It enables us to see relationships and connections that are invisible to others. Sometimes they are frivolous, like in a card trick. But in other contexts, it is vitally important – whether it is stock markets, health trends or climate predictions. Sometimes, shining a mathematical light on an otherwise dark area of human experience can be the difference between life and death!

CHAPTER 24

MATH ERROR

While we've been considering sunflowers, the Golden Ratio and the Fibonacci sequence, we've also been thinking a lot about how multiplication and division work. Several surprising patterns and characteristics emerge when you play with these quite rudimentary operations. In this chapter I want to think a little more about the idea of division and tackle one of the classic chestnuts, which I mentioned in those earlier chapters, and one that has puzzled people around the world ever since it was first stated. That old chestnut is: **why can't we divide by zero?**

Before we can unpack that conundrum, we need to take a few steps back to an earlier point in the story. A fitting place to begin is the division symbol itself, which is actually called an obelus. You've probably written and read thousands of these over your lifetime, but you might not have noticed that it visually embodies the act that it represents:

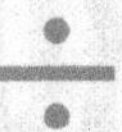

'Obelus' is the Ancient Greek word for a sharpened stick – it has the same root as the word 'obelisk' (that famously pointy piece of antiquated architecture). It indicates that what we are really focusing on is the line in the middle of the symbol, which is literally dividing the two dots into separate groups. The division symbol is a literal illustration of actual division.

The act of dividing into equal groups is the first way that we meet the idea of division as children. We have a large quantity of identical items and we want to share them out equally among a specific number of people – so we divide the identical items into groups, making sure to keep them all the same size as we go.

Say we had 24 chocolate-chip cookies and we wanted to share them between three people. This is how we could divide them:

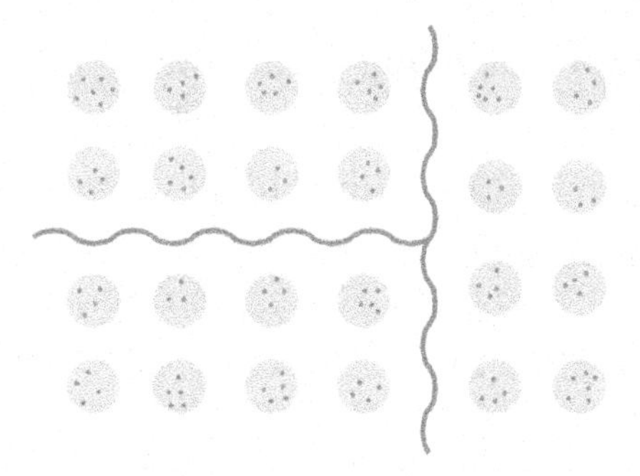

This visually demonstrates why 24 ÷ 3 = 8. What is the meaning of eight in this case? It's the number of items in each group, the number of cookies that each person gets to enjoy.

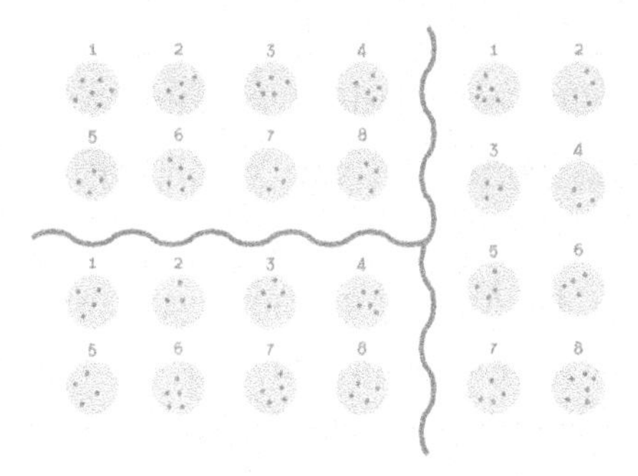

This kind of division is known as 'partition division', because you are literally taking a number and 'dividing it into parts'. The answer you get at the end (eight) is a description of the size of each equal part. But did you know that this isn't the only way that our minds understand division?

Let's start with the same 24 cookies but pose a new question. Suppose I wanted to package these to sell to people instead of just giving them away to my friends. If we want to place three cookies in each package, how many packages will we end up with? Here's what it looks like if we represent it visually:

This kind of division is called 'quotition division'. The word comes from the Latin 'quot', which translates to 'how many' (which is why a 'quota' is how many things you are required to do, and a 'quote' is how much a job is supposed to cost). In this context, the 'how many' refers to 'how many groups can you make?' This means that $24 \div 3 = 8$ can have quite a different meaning. It's the number of groups you can form, given a specific group size.

The answer is still eight. That's because what we have essentially noticed through partition is that $3 \times 8 = 24$, since we had three groups of eight cookies each, but what we observed through quotition is that $8 \times 3 = 24$, since we could make eight packages of three cookies each from the same original number. The reversibility of multiplication is a property that we call 'commutativity'.

Seeing division through the lens of quotition takes a bit more mental effort, but it can be very useful. For instance, it enables us to make sense of questions such as: what is $24 \div \frac{1}{2}$?

Answering this question through partition is possible, but weird. How are you supposed to visualise sharing your cookies . . . with half a person? Quotition, on the other hand, provides a completely natural interpretation: it means that we are making packages which each have half a cookie in them (I guess austerity measures are in force!). What we are asking is: how many packages can we make if they each contain half a cookie?

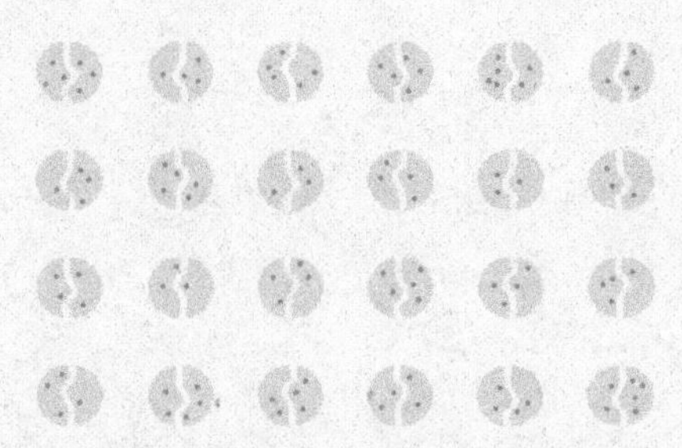

While this is sad to imagine (how is anyone supposed to ever stop at half a cookie?), it makes the question simple to answer. 24 ÷ ½ = 48.

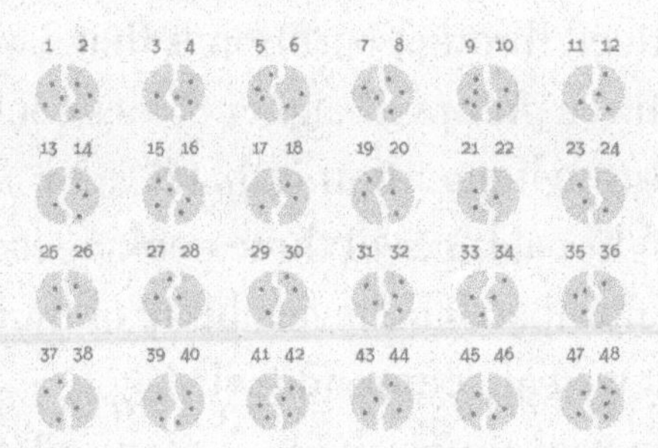

> Who knew there were so many intricacies in understanding something as apparently straightforward as division?

Now we're finally ready to return to that old chestnut I mentioned earlier – dividing by zero. It's become something of an axiom among mathematics students (not to mention their exasperated parents and teachers) to say, 'You just can't divide by zero.' Many people can tell you that this is the case, but few people have any idea as to why this should be the case. It was just an arbitrary rule that you were required to memorise at some point in your life and the sheer bizarreness of the phrase has meant that the fact has lodged itself in your mind. Some might appeal to their calculators, which will faithfully state 'MATH ERROR' when you try to divide any number by zero. That might convince some people, but at the same time it begs the question: who told the calculator to say that, and why?

The puzzle pieces we assembled earlier this chapter will help us make sense of this. Let's think about this as a partition problem. How many cookies will each person get if we share our cookies with zero people? Well, if there aren't any people to begin with, a logical conclusion you might come to is zero, since no one gets any cookies (how depressing).

But looking at the same problem from the perspective of quotition shows us that the answer is not so straightforward. How many packages can we make if each package has zero cookies in it? Well, we can make any number of packages we like – because we will never run out of cookies, seeing as we aren't putting any cookies in the packages to begin with. From this logic you could almost argue that you could make an infinite number of packages – so is the answer supposed to be infinity? This is a bit of a contradiction to the answer we came up with earlier through partition – a warning sign that indicates we are not applying logic in a consistent way.

The last nail in the coffin of this riddle comes when we think about how division relates to multiplication. Remember that $24 \div 3 = 8$, but this is because $24 = 8 \times 3$.

You can see how I get from the first equation to the second by multiplying both sides of the equation by three.

Well, taking a leaf out of the book of algebra like we did earlier for our card tricks, let's suppose that when you divide a number by zero you *do* get a reasonable answer. I don't know what it is, so let's call it x. What we are proposing is this:

$$24 \div 0 = x$$

But if we apply the same thinking that we did before, and multiply both sides by zero, we get this statement:

$$24 = x \times 0$$

While zero might be perplexing, one thing that I definitely know about it is that when you multiply it by something – anything – you will get zero. That means that there is no possible number you could substitute for x that would make a true statement.

Mathematicians say that division by zero is 'undefined'. What they mean, as we've just shown, is that division by zero is undefined because it is 'undefinable' – there is no definition you could give it that would make sense with the existing laws of mathematics. And *that* is the real reason why we can't divide by zero!

CHAPTER 25

WHY AREN'T LEFT-HANDERS EXTINCT?

My brother is left-handed. This small fact left an indelible mark on me growing up. It started with little things, like the fact that I always had to be conscious of which chair I sat in at the dinner table – because if I sat on my brother's left, we would spend the entire meal irritatingly bumping elbows. Then it made me aware of other perplexing facts, such as that if you try to cut a piece of paper with some pairs of scissors using your left hand instead of your right, the scissors fold the paper rather than cut it.

As I grew older and became interested in learning music – especially the acoustic guitar – I realised that despite looking quite symmetrical, guitars are geared towards right-handers (and yes, you can buy specially crafted left-handed guitars). My brother also taught me that the way we write Western script – from left to right – also appears to be biased towards right-handers, since left-handers are constantly smudging the words they've just written with their left hand as it rests on the paper.

My dad was also left-handed. I say 'was', in the past tense, because he grew up in a time and place where left-handedness was seen as wrong – so at school, Dad was 're-educated' to ignore his dominant hand and use his right hand for tasks like writing. Knowing all of this while I grew up made me very curious about left-handedness – in fact, I remember in primary school it even made me slightly envious because I felt very boring being an 'ordinary' right-hander. My brother, on the other hand, was under no romantic illusions about how hard it was to be left-handed;

living in a right-handed world is no fun,

he would sometimes remind me.

Little did I know that my brother and I had stumbled onto the reason why left-handedness exists in the first place. The only problem was that I didn't understand that reason until I was well into my university years. To grasp what's going on here, we need to think about a very simple idea: the survival of the fittest.

When Darwin first unleashed *On the Origin of Species* onto the scientific world, its ideas were revolutionary. We take them for granted now, so much so that we often forget how clever they are. The key insight was that the traits most likely to survive in a population – and hence be passed on to subsequent generations – are those traits that give their owners a competitive advantage. Ever wonder how so many animals have developed such stunningly effective camouflages for their habitats? The more they blended in, the less likely they were to be eaten – which meant they survived another day, and were more likely to reproduce. Their more visible brethren were easier for predators to locate and eat, removing them from the gene pool. Conversely, in other settings, strength or speed are the key characteristics that serve to protect their owners. The 'fit' survive, while the rest perish. Desirable qualities endure, while undesirable ones are relegated to the dustbin of evolutionary history.

Shy and retiring

This might seem perplexing at first, because left-handedness does not seem to be an especially desirable quality. While a good complexion and a well-toned figure are more or less universally regarded as attractive traits, few people list 'left-handed' in their Tinder profile hoping to attract a prospective partner. As my brother often pointed out, being left-handed in a world of right-handers is often just downright inconvenient. And this is, in fact, the best case scenario: John W. Santrock wrote, 'For centuries, left-handers have suffered unfair discrimination in a world designed for right-handers.'

Throughout history, left-handers have been derided by the societies in which they lived. People were treated far worse than my father – some were regarded as malicious, unlucky or even as witches for being so different. Even while most people in the world today have left behind such discriminatory practices, our languages still bear signs of this hatred today: the word for the direction 'right' also means 'correct'. The words 'dexterity' and 'dexterous' are derived from the Latin word for 'right', while 'sinister' comes from the Latin word for 'left'.

To be fair on peoples of the past, this hatred for left-handers wasn't all superstition. Warriors of many different cultures have traditionally worn their swords in a scabbard attached to their left leg, to make it easier to reach across and access with their right hand in times of need. (To this day, it is the reason why extending your right hand to someone else is a gesture of friendship and peace – because doing so means that you can't simultaneously be brandishing your sword. We don't habitually carry swords around any more, but we still shake hands!) Left-handers, on the other hand, could conceal their weapon on their right leg without drawing suspicion. This is a major plot point in the account of Ehud, an Israelite warrior recorded in the Bible's Book of Judges. His left-handed subterfuge is highlighted as the way he's able to assassinate an enemy king:

> Then the people of Israel cried out to the LORD, and the LORD raised up for them a deliverer, Ehud, the son of Gera, the Benjaminite, ***a left-handed man.*** The people of Israel sent tribute by him to Eglon the king of Moab. And Ehud made for himself a sword with two edges, a cubit in length, and ***he bound it on his right thigh under his clothes.*** And he presented the tribute to Eglon king of Moab.
>
> Ehud reached ***with his left hand, took the sword from his right thigh,*** and thrust it into [Eglon's] belly.
> (Judges 3:15–17, 21, emphasis added)

So the suspicion of ancient peoples towards left-handers was at least partly justified. But that begs the question: if cultures around the world have had such disdain for left-handers, making it a pretty undesirable quality, how can this make sense with the survival of the fittest?

How can there still be left-handers in circulation if left-handedness is such an unfavourable characteristic?

There are two pieces of information that might help us see how left-handedness has persisted through centuries of seeming undesirability. The first is to remember that in 'survival of the fittest', the meaning of 'fit' depends entirely on your context. For example, I mentioned before that camouflage is obviously a very useful quality to keep you alive and able to reproduce. But many animals have evolved the very opposite of camouflage – lavish displays of colour and shape that are intended specifically to attract attention (usually that of the opposite gender). In such cases, where standing out visually to a potential mate is the key to having children, developing awesome camouflage and blending into your surroundings would be genetic suicide rather than a shrewd strategy.

So left-handedness must be an advantage in some way. The question then becomes: what way is that? Interestingly, the sports field can provide us with some insight here.

Attention-seeker

On cricket fields and baseball pitches around the world, there are left-handers who have worked their way into some very useful niches. The reason for this is precisely because of how unusual they are. A left-handed bowler running up to the pitch looks strange: when they release the ball towards the batsman, it comes from an unusual and uncommon angle that can take the batsman by surprise. The confusion only goes in one direction, though, as the left-handed bowler has been bowling at right-handed batsmen for his entire sporting life and is entirely comfortable with facing them.

A similar phenomenon is visible in the boxing ring. Boxers do not stand symmetrically when they fight. The reason for this is simple: most boxers are not symmetrical themselves. They usually have a dominant hand, which accordingly has resulted in them having a stronger arm on that side of their body. In a fight, that makes a huge difference.

The most common boxing stance, called the orthodox (which comes from the Greek word for 'right'), sees the boxer place their left foot and left hand in front and closer to their opponent. This way, their initial hits (called jabs) come from their weaker left hand, while their stronger hits (called hooks) come from their dominant right hand. Due to the predominance of right-handers, boxers quickly develop muscle memory and become accustomed to guarding against jabs from their opponent's left hand and hooks from their opponent's right hand.

But left-handed boxers reverse this expectation, since their natural strengths lead them to stand in what essentially looks like a mirror image of the orthodox stance. This is called the southpaw stance, and it causes trouble for an opponent precisely because of its rarity: since boxers seldom encounter it, they have far less practice in fighting it and can't deal with it as competently. This effect is so strong that right-handers who are sufficiently skilled will sometimes go against their natural strength and take on a southpaw stance, just to unnerve the combatant facing them.

This may explain how left-handedness has survived.

Survival of the fittest assumes a competitive environment,

where your ability to survive and pass on your genes actually depends on your ability to fight off enemies to protect yourself

Southpaw stance

and your children. The ability to bring a secret weapon to the fight – a stance that makes you more difficult to defeat – is a handy evolutionary advantage. In fact, the rarer the stance is, the less experience others will have with fighting it, making it even more useful. If such a genetic trait were to 'catch on' and become more common in the population, it would lose its uniqueness and hence its usefulness.

The name for this kind of phenomenon is 'frequency dependence' – that is, the fewer left-handers there are in the population, the more advantage they will have. As a result, when their numbers dwindle – which normally signals the death knell for a genetic trait in a population – they exert even more power than they did before, and are able to experience an evolutionary resurgence of sorts and increase in number again. Conversely, if numbers of left-handers were ever to balloon to unusually high proportions, their competitive advantage would evaporate and their numbers would start to reduce again. Eventually we would expect the population to reach some sort of stable ratio between left- and right-handers, a situation that we would call equilibrium. That is exactly what we observe, with populations all around the world reporting very similar numbers – around 10% – of all people being left-handed.

The relevance of this idea to handedness has been shored up by two French researchers, Charlotte Faurie and Michel Raymond, who named the idea that left-handedness might be advantageous in fights the 'Fighting Hypothesis', and conducted statistical analysis to see if they could confirm it.

How do you go about testing an idea like this? This is a tricky question in and of itself – because most people think of the scientific method as being primarily based on experiments that allow us to repeatedly check whether a particular hypothesis is true or not. Will a tissue stay dry if you put it inside a cup and submerge it upside down? Try it out and see. It will probably only take you a few attempts to convince yourself of the truth. But now the question is: how could you design a similar process to confirm (or disprove) the Fighting Hypothesis? Not so easy. Faurie and Raymond decided to employ one of the most fearsome weapons in the researcher's arsenal: statistical correlation.

Even if you've never heard of probabilistic independence or correlation coefficients by their technical names, I can almost guarantee that you've interacted with these ideas before. Any time you've read a news headline that sounded something like 'Chocolate eaters live longer, say scientists' or 'People who swear are more honest', someone is smuggling statistical correlation into the conversation. Sometimes it is well-intentioned journalism, while other times it is unabashedly sensationalist. So, what is it and how does it work?

Suppose you lead a government organisation charged with improving the health of your citizens. You might decide that you want to focus on a specific health issue that is highly relevant to your country, such as obesity. In order to focus your efforts where you can make an appreciable difference, a sensible first step would be to gather data on which areas of the country have the highest rates of obesity relative to their population. Let's imagine that the table on the next page shows the kind of information you end up with.

TABLE 1

Suburb	Percentage of population classified as 'obese'
Berryville	3.5%
Shakespear Hills	3.8%
Khantown	23.0%
West Heinrichson	16.3%
North Sainsbury	8.8%
Doughtybrook	22.3%
Monks Fields	19.7%
Las Cuevas	8.1%

Once you work out which regions have the highest obesity rates, a natural question to ask next would be: why are these areas worse off than others? Working out the underlying cause that makes obesity rates high in one place compared to another might help you design effective strategies for improving the situation. Suppose you had access to a whole range of other statistics for these same regions and decided to do some comparisons. Perhaps you might be able to find a pattern.

TABLE 2

Suburb	Percentage of population classified as 'obese'	Average age of citizen	Average daily temperature	Average TVs per household
Berryville	3.5%	39.4	26.5	0.9
Shakespear Hills	3.8%	36.1	28.1	1.2
Khantown	23.0%	34.7	26.4	6.4
West Heinrichson	16.3%	32.3	27.2	4.4
North Sainsbury	8.8%	37.3	23.5	2.7
Doughtybrook	22.3%	35.6	25.0	6.0
Monks Fields	19.7%	31.0	22.7	5.1
Las Cuevas	8.1%	30.9	27.4	2.4

Nothing jump out at you just yet? That's okay – this data looks pretty haphazard right now because it hasn't been sorted into any kind of useful order. Let's try again, but this time we'll rearrange the data a little in Table 3.

TABLE 3

Suburb	Percentage of population classified as 'obese'	Average age of citizen	Average daily temperature	Average TVs per household
Khantown	23.0%	34.7	26.4	6.4
Doughtybrook	22.3%	35.6	25.0	6.0
Monks Fields	19.7%	31.0	22.7	5.1
West Heinrichson	16.3%	32.3	27.2	4.4
North Sainsbury	8.8%	37.3	23.5	2.7
Las Cuevas	8.1%	30.9	27.4	2.4
Shakespear Hills	3.8%	36.1	28.1	1.2
Berryville	3.5%	39.4	26.5	0.9

All the same data is here, but we've ordered the list of suburbs according to their obesity rate. In other words, we've ranked them from highest to lowest. The reason this is useful is because that is the factor we're interested in understanding, so if there is a pattern in any of the other pieces of information that matches the ranking we see in the obesity rate, that gives us an indication that something could be worth investigating further.

Statisticians and other people who work with information call this process 'analysis' – so if you ever meet someone who calls themselves a 'data analyst', you can know that part of what they do is take huge heaps of messy data and try to arrange them in ways that help us see the underlying structures that aren't obvious at first glance.

Now's the perfect time to introduce you to another instrument in the statistical toolbox: visualisation. Human thinking is dominated by sight, with four times as much brain matter devoted to visual processing than to touch, and ten times as much as hearing. Almost half of all neural tissue deals with vision in some way, more than all the other senses combined. That's why it makes sense that we are better at understanding things when they come to us in the form of pictures.

There are literally hundreds of ways that we could visualise this data, but let's use a staple from the statistician's cookbook: the scatter plot. As the name suggests, it shows information as a series of dots scattered on a two-dimensional plane.

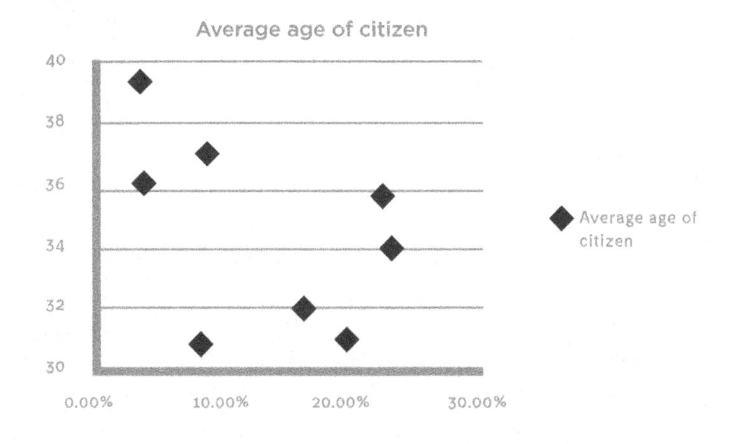

What are we looking at here? Each of the dots represents a suburb. As we move across the graph from left to right, we are going from the lowest obesity rate to the highest. The higher the dot is, the higher the average age of the citizens in that suburb.

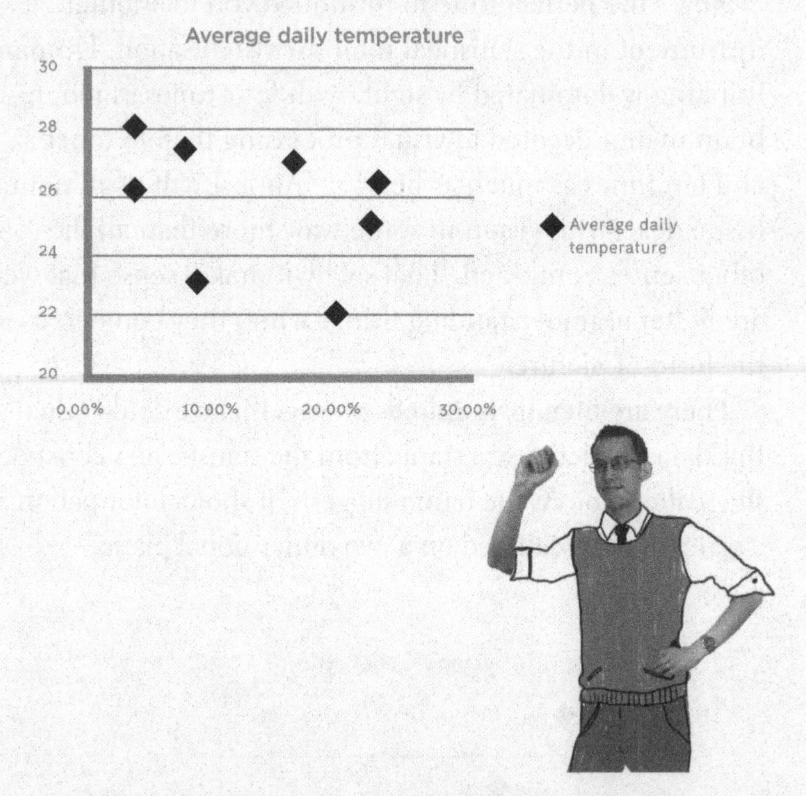

Here's the same thing, except we're now comparing obesity rates with climate. Do higher obesity rates match up in any way with warmer or cooler temperatures? There doesn't seem to be any kind of visible pattern or relationship between these two ideas either.

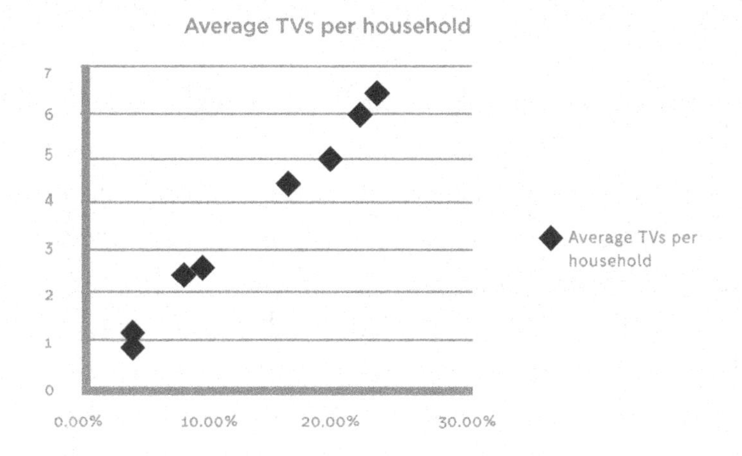

But now look at what happens when we compare the obesity rate with the average number of TVs per household. A clear shape emerges in the data. Lower rates of obesity seem to match up with low numbers for TV ownership, and the opposite is true on the other end of the spectrum. There also seems to be matching behaviour in the middle of the graph, where an increase in numbers of TVs appears to correspond with a rise in the obesity rate. This is called statistical correlation.

You can picture the headline now. 'TVs cause obesity!' But here is where it is important to recognise that just like any tool, statistical correlation can be misused. The saying that captures this is 'correlation does not imply causation' – that is, just because two quantities rise and fall in proportion to each other does not mean that one is the direct cause of the other. The link may be completely coincidental. Or, as is more likely the case in this scenario, there may be some underlying factor that is the real cause driving change in both of the factors we are investigating.

For this situation, a reasonable hypothesis would be that increasing household income leads to higher rates of obesity (since it enables people to purchase more of the foods likely to lead to obesity), and that it simultaneously leads to people purchasing more televisions for their homes (since they have the money to do so).

Which brings us back to our troublesome left-handers. Faurie and Raymond wanted to see if there was some kind of connection between left-handedness and fighting prowess. What kind of data did they look at to see if there was a pattern? What kind of cultures might have different rates of left-handedness, and what might demonstrate the importance of fighting skill? In their study, they decided to compare eight traditional societies for their homicide rate. They recognised that looking at modern cultures would skew their results since fighting skill is less pronounced as a desirable trait for potential partners. And the scatter plot below shows what they found.

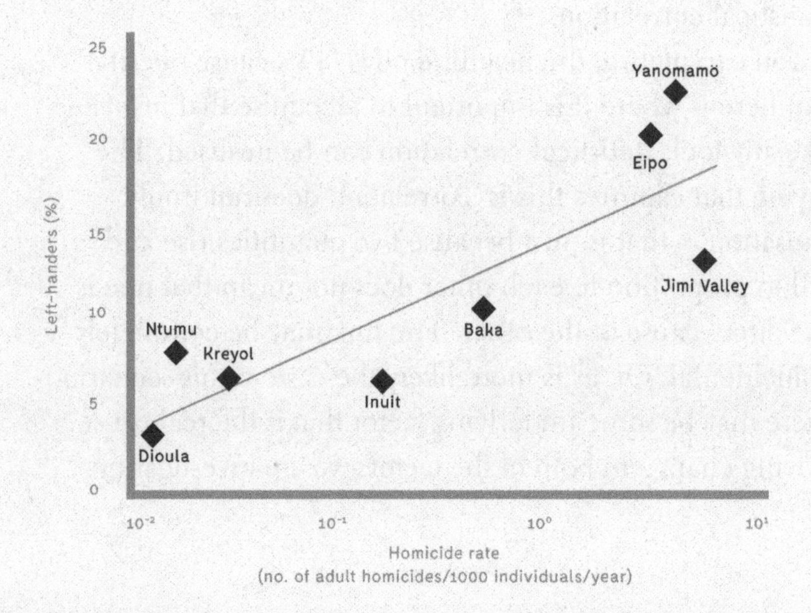

The correlation is certainly there! So does this mean that the fighting hypothesis has been proven? Well, no. Far from it, in fact – as we showed with our illustration of obesity rates and TV ownership, there might be much more lying underneath the surface. Or this relationship could even be coincidental. But one thing is certain: this is an additional piece of evidence to back up what mathematical logic tells us about the usefulness of a trait like left-handedness.

CHAPTER 26

WHY YOUR PANCREAS IS LIKE A PENDULUM

In the previous chapter, while thinking about why left-handers haven't disappeared from the genetic pool, I briefly mentioned the idea of equilibrium. This is a common idea across the sciences, since many situations that we observe in nature deal with competing forces that have arrived at a state of balance. This makes sense, because any situations that do not eventually establish balance will, by definition, cease to exist after an extended period of time.

The concept of competing forces is an interesting one to consider mathematically because it often creates a pattern like this:

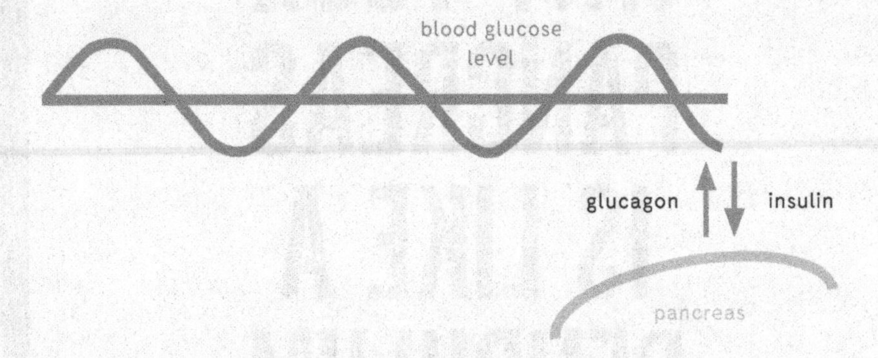

This graph represents the blood glucose level over time, showing the way that it is moderated by the pancreas. The straight line is called the 'homeostatic set point', which you can think of as the 'normal' amount of glucose in your bloodstream at any given time. Elevated blood sugars are a big problem for the human body, and hyperglycaemia can cause seizures or even death in extreme cases. When the pancreas senses glucose levels rising too high, it releases insulin to slow and eventually stop the increase.

The opposite situation, hypoglycaemia, is equally dangerous, and so the pancreas avoids it by releasing glucagon to restore the balance. This is why people with diabetes, whose pancreas does not function normally, often carry around sweet lollies such as jelly babies to eat in emergency cases where the body's blood sugar level suddenly drops. A very similar mechanism, called the baroreflex, works in the same way to sensitively moderate your body's blood pressure.

But you might recognise the characteristic shape of the glucose level graph from the earlier chapter, 'Music to my ears'. That's right – it's a sinusoidal wave, the same kind that makes up musical notes.

These wave patterns arise in systems that have something called 'negative feedback'. Many mechanical systems are designed with negative feedback in mind, and as you can see in the diagram, many biological systems also function in this way. Whenever we want the maintenance or regulation of some environment or condition, negative feedback is the tool for the job.

You can create your own version of this to observe the usefulness of negative feedback by taking a length of string or anything similar – dental floss, a shoelace, or whatever else is on hand – and then tying something heavy onto one end. Let the heavy object hang down under the force of gravity, and then swing it in any direction you like. Now stop moving your hand and observe what happens. You've just created one of the simplest negative feedback systems we know of: the pendulum.

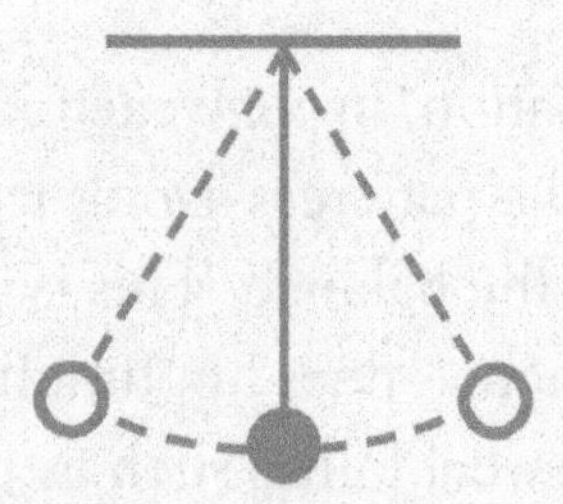

We've been using the regular motion of pendulums to tell time since the seventeenth century. Negative feedback is at the heart of how the pendulum works: once it swings too far in one direction, its distance away from the centre sends it hurtling back towards the other side. And hence the motion can continue with very little intervention. And what happens if you trace out the path of the pendulum bob over time as it swings from side to side?

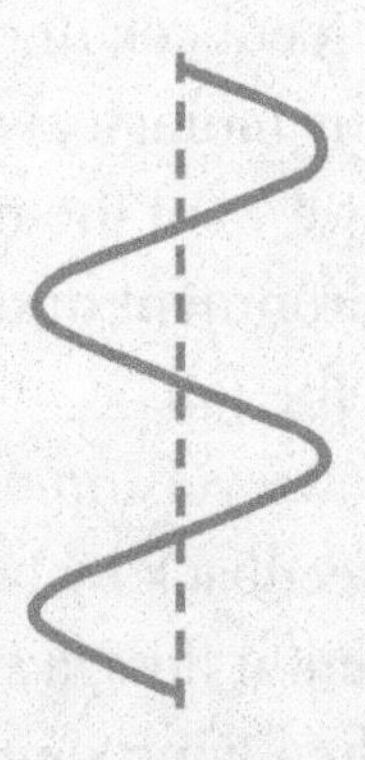

You guessed it – our old friend, the **sinusoidal wave!**

Negative feedback can be observed in economic situations, too. Consider the following graph that shows the history of real estate prices in Australia since 1926:

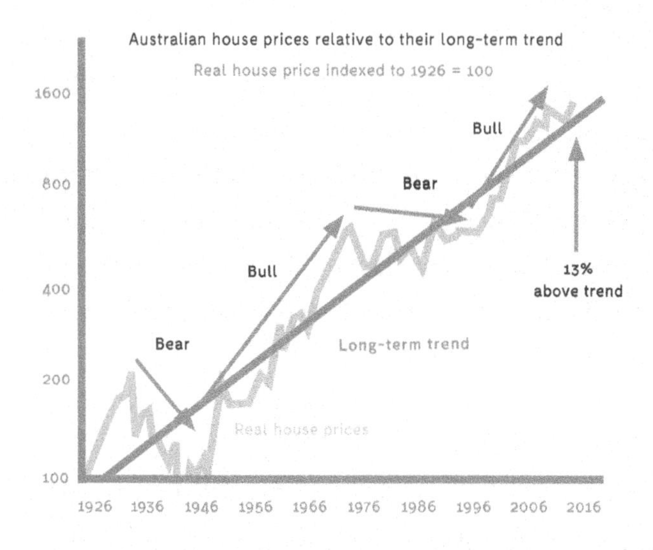

The entire graph does trend upwards, which reflects things such as inflation and gentrification. However, if you ignore this overall shape, the underlying sinusoidal pattern is there. This reflects the negative feedback loop that comes from the economic cycle of supply and demand. As properties are purchased and become occupied, their increasing rarity leads to elevated demand and higher prices (real estate pundits call this a 'bull market'). In response to this shortage, though, the government will re-zone land for development and dramatically increase the number of available properties, reducing demand and driving prices back down (a so-called 'bear market').

Just like we observed when thinking about fractals (see the earlier chapter, 'Lightning through your veins'), this is one of those wonderful places where mathematics helps us see that sometimes a single idea or structure may be hiding underneath things that seem completely different on the outside.

This also sheds light on why so many people – perhaps including you – experience difficulty in their mathematical learning journey. By its very nature, mathematics focuses on those underlying ideas that relate to everything, with all the specific details and context removed. When we see symbols such as 'x^2', we know it means 'multiply x by itself' – but x could be a length of a triangle, or an amount of money, or the speed of light.

However, taking those details and context away often makes ideas harder for our brains to get a handle on. Not only can it make things harder to comprehend, it also takes away some of the reasons for why we would care about these symbols at all – why should I be interested in x if I don't know what it represents? And these two factors alone account for the lion's share of why people experience algebra as such a difficult topic to work with.

This is not a shortcoming or design flaw, though. It's precisely what makes mathematics so powerful and useful.

Mathematics is the ultimate skeleton key: if you can learn to wield it, it can unlock almost anything.

FURTHER READING

Acheson, David, *1089 and All That*, Oxford University Press, Oxford, 2002.

Bellos, Alex, *Alex's Adventures in Numberland*, Bloomsbury Publishing, London, 2010.

Devlin, Keith, *Mathematics: The Science of Patterns*, Henry Holt and Company, New York, 1994.

du Sautoy, Marcus, *The Number Mysteries: A Mathematical Odyssey Through Everyday Life*, St Martin's Press, London, 2011.

Fry, Hannah, *The Mathematics of Love*, Simon & Schuster Inc, New York, 2015.

Parker, Matt, *Things to Make and Do in the Fourth Dimension*, Penguin Books, 2014.

Singh, Simon, *The Simpsons and Their Mathematical Secrets*, Bloomsbury Publishing, London, 2013.

Spencer, Adam, *Adam Spencer's Big Book of Numbers*, Xou Pty Limited, Sydney, 2014.

Strogatz, Steven, *The Joy of X*, Atlantic Books, London, 2012.

ACKNOWLEDGEMENTS

Even if there is only a single name on the front cover of a book, no work of this size or quality is the product of a solitary individual. This book is no exception – a whole team of people has helped make this dream a reality.

Claire Craig believed that I had a book in me, even before I did. Thank you for reawakening my love of writing. I thought that love had been effectively buried many years ago – but you helped me see that it was just taking a nap.

Rebecca Hamilton and Brianne Collins combed through my thousands of words with admirable patience to sharpen every sentence (and every diagram!). Special thanks to Bec for putting your money where your mouth is and bravely trying out a mathematical card trick in front of a live audience!

Alissa Dinallo brought these pages to life with her artistry. This book was always going to be intensely visual from the very beginning, so thank you for putting so much time into understanding the concepts and partnering with me to help people sense these realities!

Dylan Wiliam and Keith Devlin provided invaluable mathematical advice. Everything I've written is so much the better for your constructive thoughts as 'capital M' Mathematicians.

Jenelle Seaman helped to sharpen my scientific knowledge when I was roaming into the areas of biology and chemistry – thank you from one educator to another.

Lastly, my deepest thanks must go to my family. Michelle – I'm so grateful that you put up with absolutely every aspect of my crazy personality and love me anyway, especially in this most recent season of life that has been such a roller-coaster. Emily, Nathan and Jamie – thank you for filling my life with such joy, and also for giving me a new excuse to fall in love with books all over again!